To Malcolm James, A.E. 'Jimmy' James, and the Wandering Islands

and

To Gimpock P. Lew and Inge Ida Lew

Contents

Figures

Tables

Contributors

Christine Butalla	Comstock Park, Michigan, USA.
Richard Butler	School of Management Studies, University of Surrey, Guildford, Surrey, UK.
Clive Charlton	Department of Geographical Sciences, University of Plymouth, Plymouth, Devon, UK.
Bill Forbes	USDA Forest Service, Brookings, Oregon, USA.
C. Michael Hall	Centre for Tourism, University of Otago, Dunedin, New Zealand; Senior Research Fellow, New Zealand Natural Heritage Foundation; and Visiting Professor, School of Food and Leisure Management, Sheffield Hallam University, Sheffield, UK.
Thomas D. Hinch	Faculty of Physical Education and Recreation, University of Alberta, Edmonton, Alberta, Canada.
John Hull	Department of Geography, McGill University, Montreal, Quebec, Canada.
Alan A. Lew	Department of Geography and Public Planning, Flagstaff, Arizona, USA, and Visiting Fellow, Department of Geography, National University of Singapore.
Simon S. Milne	Department of Geography, McGill University, Montreal, Quebec, Canada and Auckland Institute of Technology, Auckland, New Zealand.
Stephen Page	Tourism, Department of Management Systems, Massey University – Albany, New Zealand.
Susan E. Place	Geography and Planning and Latin American Studies, California State University, Chico, California, USA.
Matt Pobocik	Comstock Park, Michigan, USA.
Gareth Shaw	Tourism Research Group, Department of Geography, University of Exeter, Exeter, UK.
Kaye Thorn	Tourism, Department of Management Systems, Massey University – Albany, New Zealand.
Alan Williams	Tourism Research Group, Department of Geography, University of Exeter, Exeter, UK.

Pamela Wight   Pam Wight & Associates Inc., 14715–82 Avenue,
                    Edmonton, Alberta, Canada.

Heather Zeppel   Leisure Studies, University of Newcastle, New South
                    Wales, Australia.

Acknowledgements

Geographers, and tourism geographers in particular, have long had interests in sustainable development and natural resource management. However, this particular book had its genesis in the activities of the Association of American Geographers Recreation, Sports and Tourism Study Group and the International Geographical Union Study Group on the Geography of Sustainable Tourism and their wide-ranging interests in sustainable tourism development. Both organisations have been extremely supportive in the development of this particular project, which aims to convey both a theoretical and an empirical snapshot of the activities of geographers in the field of sustainable tourism development.

We would like to thank Matthew Smith and Addison Wesley Longman in the United Kingdom for their encouragement in the development of this book and their overall support for publishing in the field of tourism and recreation geography. Support has also come from our various academic departments over the past two years at the National University of Singapore, North Arizona University, the University of Otago and Victoria University of Wellington. Angela Elvey, Kirsten Short and Janine Watkin provided much-appreciated assistance in the preparation of the manuscript, while acknowledgement is also due to Shawn Colvin, Dave Crag, Neil and Tim Finn, Linda Kell, Sara McLachan, Natalie Merchant, Kirsten Short and Chris Wilson for their earlier help. Finally, we would like to thank our family and partners for their continued love and support.

C. Michael Hall   Alan A. Lew
Dunedin   Singapore

Chapter 1

The geography of sustainable tourism development: an introduction

C. Michael Hall and Alan A. Lew

Defining and achieving sustainable development has become one of the major policy debates of our generation. While concerns over the use of natural resources and their relationship to economic growth have been a significant issue for governments in Western countries since at least the late nineteenth century, at no time have such issues been so high on local, national and international policy agendas for so long.

Debate over the 'wise use' of natural resources has been at the centre of geographical imagination for many years. George Perkins Marsh's book *Man and Nature or, Physical Geography as Modified by Human Action* (1965), originally published in 1864, had enormous impact on conservation debates, the effects of which are still reverberating to the present day. Since the time of Marsh, geographers have been influencing the course of natural resource management in several ways (see Mitchell 1989):

- in terms of policy advice to governments and business;
- in the establishment of frameworks and methodologies for the analysis of human impact and interaction with the physical and social environment; and
- in charting the relationship between society, individuals and the environment.

While the field of resource management is sometimes regarded as a specific sub-field of applied geography, it is worth noting that all of the sub-disciplines and traditions of geography have studied the human–environment relationship in its various forms. Indeed, the realisation not only by geographers, but also by other academic disciplines, governments, industry and, increasingly, the wider public, that environmental and economic issues present complex systemic problems that require appropriate holistic solutions, has served to provide a renewed relevance for the geographer's craft.

The present debate over issues of sustainability and sustainable development represents the current manifestation of recognition in certain parts of Western society of an 'environmental crisis' that has been with us to varying degrees since the time of Marsh. Although Chapter 2 by Hall will provide further discussion of the historical roots of sustainability, it is valuable to identify some of the key elements of the definition of sustainable development and the manner in which it influences current debate before going on to examine some of the key aspects of sustainable tourism.

The concept of sustainable development

The concept of sustainability first came to public attention with the publication of the World Conservation Strategy (WCS) in March 1980 (IUCN 1980). The WCS was prepared by the International Union for Conservation of Nature and Natural Resources (IUCN) with the assistance of the United Nations Environment Education Programme (UNEP), the World Wildlife Fund (WWF), the Food and Agriculture Organisation of the United Nations (FAO) and the United Nations Educational, Scientific and Cultural Organisation (UNESCO). The WCS was a strategy for the conservation of the Earth's living resources in the face of major international environmental problems such as deforestation, desertification, ecosystem degradation and destruction, extinction of species and loss of genetic diversity, loss of cropland, pollution, and soil erosion. The WCS was developed by a combination of government agencies, non-government organisations and individual experts from over 100 countries.

The WCS defined conservation as 'the management of human use of the biosphere so that it may yield the greatest sustainable benefit to present generations while maintaining its potential to meet the needs and aspirations of future generations' (IUCN 1980: s.1.6). The WCS had three specific objectives (*ibid.*: s.1.7):

1 to maintain essential ecological processes and life-support systems (such as soil regeneration and protection, the recycling of nutrients, and the cleansing of waters), on which human survival and development depend;
2 to preserve genetic diversity (the range of genetic material found in the world's organisms), on which depend the breeding programmes necessary for the protection and improvement of cultivated plants and domesticated animals, as well as much scientific advance, technical innovation, and the security of the many industries that use living resources;
3 to ensure the sustainable utilization of species and ecosystems (notably fish and other wildlife, forest and grazing lands), which support millions of rural communities as well as major industries.

The notion of sustainable development espoused in the WCS emphasised the relationship between economic development and the conservation and sustenance of natural resources. In many ways, there was nothing new in this idea as it had been at the core of much of the conservation debate for many years (see Chapter 2). However, what was significant was the manner in which the report highlighted the global nature of environmental problems, emphasised the significance of the environmental–economic development relationship in the relationship between the developed and less developed countries (the North–South debate), and provided a basis for some government and private sector response, albeit limited, to the problems and issues identified in the report.

The WCS was also significant in that it represented the halfway mark between the 1972 United Nations Stockholm Conference on the Human Environment and the 1992 UN Conference on Environment and Development (UNCED) in Rio de Janeiro (often referred to as 'the Rio Summit'). In a very real sense, the WCS can be regarded as a belated child of the Stockholm Conference, as one of the institutional results of the meeting was the establishment of the UNEP, which had substantial influence on the formulation of the WCS. In addition to assisting in the development and promotion of the WCS, the UNEP promoted the idea of the creation of a World Commission on Environment and Development (WCED) at its ten-year review conference in 1982. In 1983, the Commission was created as an independent commission reporting directly to the United Nations Assembly, with Gro Harlem Brundtland, then parliamentary leader of the Norwegian Labour Party, being appointed as its chair. Although the term 'sustainability' was also used in a 1981 book by Lester Brown of the Worldwatch Institute, *Building a Sustainable Society*, by Myers (1984) and Clark and Munn (1986), *Ecologically Sustainable Development of the Biosphere*, it was not until the publication of the report of the WCED in 1987, *Our Common Future*, commonly referred to as the Brundtland Report, that sustainable development entered into the popular lexicon.

According to the WCED (1987: 43) sustainable development is development that 'meets the needs of the present without compromising the ability of future generations to meet their own needs'. Five basic principles of sustainability were identified in the report.

1 the idea of holistic planning and strategy making;
2 the importance of preserving essential ecological processes;
3 the need to protect both human heritage and biodiversity;
4 to develop in such a way that productivity can be sustained over the long term for future generations; and
5 achieving a better balance of fairness and opportunity between nations.

Nevertheless, despite the promise and essential appeal of the concept of sustainable development, considerable uncertainty exists over its definition (e.g. Robinson *et al.* 1990), particularly with respect to the relationship of sustainability to economic growth (e.g. Redclift 1987; Sachs 1993), and the manner in which it is implemented and operationalised (e.g. Nath, Hens and Devuyst 1996). Furthermore, achievement of sustainable development may also require a change in lifestyle in Western countries, not only to place less pressure on the use of non-renewable resources but also to assist in the transfer of funds from the North to the South (WCED 1987). The concept of sustainable development therefore provides a substantial challenge to the people and governments of Western nations.

The scope of the challenge that needs to be met in order for economic activity to be environmentally sustainable has been outlined by Ekins (1993b), who argues that certain conditions need to be rigorously adhered to with respect to resource use, pollution and environmental impacts:

- destabilisation of global environmental features such as climate patterns and the ozone layer must be prevented;
- important ecosystems and ecological features must receive absolute protection in order to maintain biological diversity;
- renewable resources must be maintained with sustainable harvesting measures rigorously enforced;
- non-renewable resources must be used as intensively as possible;
- depletion of non-renewable resources should proceed on the basis of maintaining minimum life expectancies of such resources, at which level consumption should be matched by new discoveries of these resources and technological innovation;
- emissions into the biosphere should not exceed the biosphere's capacity to absorb such emissions; and
- risks of life-damaging events from human activity, e.g. nuclear power generation, must be kept at a very low level.

Clearly, meeting such conditions for sustainability is a major political, economic and environmental issue as it requires new ways of thinking about the nature and purpose of development and growth, and the role of individuals, government and the private sector in developing sustainable futures, a concern that is increasingly at the forefront in the analysis of tourism.

The geography of tourism

Although geographers have long held an interest in issues of the conservation and use of natural resources, interest in tourism and recreation did not emerge until the 1930s (e.g. McMurray 1930; Jones 1933) and, as a distinct sub-field of geography, until the late 1960s and early 1970s (e.g. Murphy 1963; Mitchell 1969; Mercer 1970; Stansfield and Rickert 1970; Lavery 1971; Cosgrove and Jackson 1972; Robinson 1976). The significance of the area is also reflected in the regular publication of reviews of tourism, leisure and recreation in *Progress in Human Geography* and individual contributions throughout the international geographical journals, particularly *Annals of the Association of American Geographers, Applied Geography, Area, Geoforum, GeoJournal* and the *Professional Geographer*, as well as various tourism journals such as *Annals of Tourism Research, Tourism Management* and the *Journal of Sustainable Tourism*. However, despite the substantial growth in the number of publications by geographers on tourism and the establishment of tourism special interest groups in such academic organisations as the International Geographical Union, the Association of American Geographers and the Canadian Association of Geographers, the study of the geography of tourism remains highly fragmented (Pearce 1979; Mitchell and Murphy 1991; Page and Hall 1998).

The study of the geography of tourism draws on a number of fields within geography. Yet, while it is quite easy to agree with Matley's (1976: 5) observation that 'There is scarcely an aspect of tourism which does not have some geographical implications and there are few branches of geography which do not have some contribution to make to the study of the phenomenon of tourism,' one must also note that the relative influence of these branches has proven to be highly variable over the past 70 years.

Mitchell and Murphy (1991) identify four major contributions of geographers to the study of tourism with respect to environmental, regional, spatial and evolutionary considerations. Undoubtedly, these four areas are of substantial significance to geographers. Questions of the relationship between tourism and the physical and social environment, particularly with respect to such notions as carrying capacity, have been at the forefront of much geographical research (Mathieson and Wall 1982; Pearce 1989). Regional studies, long a major geographical tradition, have received considerable attention in recent years, with major surveys being undertaken by geographers on Western Europe (Williams and Shaw 1988), Eastern Europe (Hall 1991), polar regions (Hall and Johnston 1995), Australia (C.M. Hall 1995), China (Lew and Yu 1995), the South Pacific (Hall and Page 1996), and the Pacific Rim (Hall 1997). Spatial considerations have also been a major focus, particularly with respect to the supply and demand aspects of tourism, patterns of movements and flows, and models of tourist space (Pearce 1979). Pearce (1991) provides a very good overview of the spatial approach to the geography of tourism and recreation. Of the four approaches identified by Mitchell and Murphy (1991), the evolutionary tradition is probably the least developed, although the attention given to the question of how tourist destinations change through time offers considerable opportunity to integrate various approaches to the study of the geography of tourism. Much of the work undertaken on resort and destination evolution focuses on the work by Butler (1980), who, in turn, had built on the earlier work of Christaller (1963).

In addition to the contributions identified by Mitchell and Murphy (1991), several other significant research fields can be identified. For example, several geographers have contributed work on tourism planning, which has been heavily influenced by the urban and regional planning tradition within geography (e.g. Murphy 1985; Getz 1986; Hall, Jenkins and Kearsley 1997), while several texts have been produced by geographers in recent years on urban (e.g. Law 1993; Page 1995) and rural (e.g. Butler, Hall and Jenkins 1998) tourism issues. Perhaps, most significantly, some of the more recent research on tourism by geographers is starting to reflect some of the wider debates in the discipline about the processes of modernisation and development, issues of gender and identity, place marketing and promotion, and the relationship between globalisation and economic and cultural change (e.g. Britton 1991; Hall 1994a; Kinnaird and Hall 1994; Shaw and Williams 1994).

That tourism geographers have begun to broaden the scope of their attention should come as no surprise. As Johnston (1991: 1) recognised, academic life 'is not a closed system but rather is open to the influences and commands of the wider society which encompasses it'. The ideas and the structure of the geography, and of the geography of tourism, 'developed in response to complex social, economic, ideological and intellectual stimuli' (Stoddart 1981: 1). Researchers in the geography of tourism are a subcommunity within the community of academics and researchers, which is itself a subset of wider society; that society has a culture, including an academic/scientific subculture, within which the content of the geography of tourism is defined. Action is predicated on the structure of society and its knowledge base: research praxis is part of that programme of action, and includes tourism research. The community of tourism geography academics and researchers is an 'institutionalizing social group' (Grano 1981: 26), a context within which individuals are socialised and that defines the internal goals of their subject area in the context of the external structures within which they operate (after Johnston 1991: 277).

Recent developments in the study of the geography of tourism noted above are not only representative of shifts within geography but also reflect wider concerns within the social sciences, which have seen a focus on various aspects of economic and cultural globalisation and localisation processes in particular (see Johnston 1991). In addition, it should be noted that the growth in the number of geographers who are actively researching and teaching in the area of tourism and recreation is also a reflection of the extent to which governments in Western countries have come to recognise the economic development and employment possibilities of tourism and accordingly increase the policy profile of the area. Indeed, the growth of government interest in tourism and the subsequent boost this has given to the legitimacy of geographical research is similar to that experienced in other disciplinary studies of tourism, such as anthropology and politics, where, until recently, tourism was seen as a topic that was thought of as being unworthy of consideration by the serious scholar (Smith 1977; Hall 1994a). It is therefore in this context that geographers are now turning to examine tourism within the wider framework of sustainable development.

Although the topic of tourism sustainabity is not the sole domain of geography (see Bramwell and Lane 1993), geographers have made considerable contributions to the definition and analysis of the field in recent years (e.g. Pigram 1990; Butler 1991; Cater 1993; McKercher 1993a, b; Wight 1993a). Similarly, research conferences, such as the Association of American Geographers annual conferences, have featured a number of well-attended sessions on sustainable tourism, while the International Geographical Union has a study group on the geography of sustainable tourism that has a substantial European, as well as North American and Australasian, membership. Yet, despite the considerable amount of research, along with student and industry interest, in geographical perspectives on sustainable tourism, no book has yet brought together some of the main strands of the geographer's focus on sustainable tourism development. This book, therefore, aims to correct this situation by providing a ready guide to current approaches, issues and experiences in geographical perspectives of sustainable tourism.

Outline of the book

The book is divided into two main sections. Following the introduction to the book, the first section examines a variety of contemporary approaches to sustainable tourism from a number of different disciplinary and sub-disciplinary perspectives. Hence, contributions are made from the fields of economic geography and cultural geography as well as the more traditional 'resource management' approaches. Such a collection of chapters helps to convey to the reader how issues of sustainability are related to contemporary geographical debates over restructuring, post-Fordism, cultural identity and place promotion, as well as research on management frameworks and techniques to ameliorate impacts on the physical and social environment. The second section of the book presents a series of case studies and sets of experiences on sustainable tourism, which vary in their location and development context. These studies have been chosen because of the manner in which they reinforce the themes and issues identified in the first section.

Approaches and issues

Chapter 2 by Hall on 'Historical antecedents of sustainable development and ecotourism' provides an historical and political analysis of concepts of sustainable development with respect to tourism and provides an historical framework of the development of the concept of sustainability for the chapters that follow. It observes that the present attempts to find a balance between economy and environment have been a central component of natural resource management since the writings of John Perkins Marsh in the 1860s. With reference to debates over natural area usage, the chapter argues that tourism has long been a component in natural resource utilisation, particularly with respect to the role of national parks and other conservation areas. The chapter charts the changing nature of conservation strategies and argues that the focus by many environmental groups around the world on national parks in the overall political debate surrounding the environment may actually be hindering achievement of processes of sustainable tourism development. The chapter concludes by observing that national parks, as with all aspects of the tourism–resources management relationship, need to be perceived within a continuum of resource use that seeks to develop integrated tourism management strategies that relate to specific communities of interest.

As with many other forms of economic activity, tourism has become inextricably linked with the concept of sustainable development, and, as noted above, much attention has been paid to 'sustainable tourism'. At the same time, the rise in environmental consciousness has seen the parallel appearance of a large number of 'alternative' forms of tourism, along with a great deal of ideological and polemical debate about the role of tourism in development and its impacts on societies, economies and environments.

Chapter 3 by Butler documents, interprets and synthesises some of the literature relevant to these trends and goes on to identify some common themes and concepts that transcend what has tended to become an often disparate and even divisive discussion. It is argued that further discussion of concepts and definitions is not likely to be of great benefit, and that attention should now shift to translating theoretical constructs into applied methodologies. In this vein, the chapter examines two specific elements of sustainability in the context of tourism, the sustainability of environments in which tourism is present or proposed, and the sustainability of societies and cultures in which tourism is present or proposed. The chapter raises and comments on the state of the art of knowledge and research on these issues through a review of the literature. The discussion addresses whether the concept of sustainability is the same in both environmental and social settings; whether it is possible to achieve sustainability without imposing limits on use and/or numbers; how technology and infrastructure development relate to sustainable development; how local values and external values can be reconciled with respect to defining and implementing sustainable development in specific locations; and how other elements of the tourism system such as local cultural and natural heritage, the economic benefits of tourism, and the tourism market of specific destinations can be sustained. In this context, particular attention is paid to three overriding features: carrying capacity, control and mass tourism. Carrying capacity, Butler argues, is at the core of sustainable development, and must be operationalised if sustainability is to be achieved. Control of tourism is related to this issue, since the rate, nature and level of development, determined through policy formulation and implementation, have crucial implications for the capacity of a destination to absorb tourism. Finally, the issue of mass tourism is addressed, since most of the world's tourism is of the 'mass' variety. To make any real progress on achieving sustainable development of tourism, the problems of mass tourism must be examined and the processes related to them made more sustainable. This, it is argued, is the real challenge of sustainable tourism but represents a challenge that has not yet been taken up by industry, government or the research community.

Chapter 4 by Milne takes up particular questions of the global–local nexus in the relationship between tourism and sustainable development and argues that much of the existing literature on tourism and development (sustainable or otherwise) is highly descriptive in nature and has been weakened by a continued fascination with theoretical models that either predict linear paths of destination evolution or speak of radical shifts from an 'old' to a 'new' tourism. Drawing on empirical research, recent debates surrounding regulation theory and the burgeoning literature on sustainable rural systems, this chapter presents a more rigorous and industry-specific theoretical framework. The framework stresses that tourism is characterised by processes of both continuity and change and that researchers must grapple more effectively with the interactions that occur both within and between various scales of analysis (global, regional and local). It also argues that researchers must look carefully at definitions of sustainability and community and in particular re-evaluate the role that communities (both urban and rural) can play in influencing the course of tourism development.

Chapter 5 by Williams and Shaw also draws on the debates that are occurring in geography over issues of economic restructuring and sustainability within a global environment. The chapter argues that the debate on sustainable tourism would be enriched by strengthening its links to the wider debates on economic restructuring. The more micro-scale concerns of, and/or advocacy role of, much of the debate on sustainable developments can be usefully placed in the context of some of the wider processes that are shaping the reorganisation of the production and consumption of tourism. The chapter therefore provides a review of four significant themes that draw on the contemporary European experiences of tourism development: shifts in the mode of production, new forms of consumption, the organisation of production, and the limits to state intervention.

Within wider debates over the relationship between tourism and globalisation processes, considerable attention has been given to the role of tourism in reinforcing or destabilising cultural identity, particularly with respect to indigenous peoples. Chapter 6 by Zeppel examines indigenous cultural experiences and their sustainability and notes that the opening up of indigenous homelands areas and wider involvement in tourism involves a balance between tourist needs for personal contact with indigenous people and maintaining environmental, social and cultural integrity. The chapter examines indigenous involvement in 'native-owned' tourism ventures in Australia, New Zealand and Canada and compares the approaches adopted by Aboriginal communities, Maori groups and First Nations peoples in developing and implementing sustainable tourism. Key aspects include indigenous values and the dilemma in achieving cultural, environmental and economic sustainability in tourism. Zeppel argues that sustainable indigenous tourism requires a balance between three variables: resilience of cultural integrity and social cohesion, maintenance of the physical environment, and commercial success. A further contrast is made between invited or community-controlled indigenous tourism ventures in homeland areas and indigenous responses to imposed tourism in national park areas now under joint management. Sustainable tourism is therefore regarded as deriving from indigenous management and control of land and resources.

Chapter 7 by Wight examines some the techniques that may be applied to assessing sustainability and falls within the applied tradition of geography. The chapter examines approaches to managing the environmental impact of recreation and tourism from a socio-economic as well as an environmental perspective. Approaches that have been advocated include carrying capacity, limits of acceptable change (LAC), visitor impact management (VIM), visitor experience and resource planning, and environmental impact assessment (EIA) and cumulative effects assessment. Carrying capacity, also noted by Butler in Chapter 3, is regarded as being useful as a concept, but there are tremendous difficulties in translating the concept to practical application. LAC does not focus on *amount* of visitation and activity but rather on the change to the environment that we are prepared to tolerate. It is a tool designed to identify appropriate or desirable resource and social environmental conditions in a given area, and to guide the development of management techniques to achieve and protect those conditions. Wight argues that the principles, rather than the specifics of the process, are of greatest relevance, introduces a values perspective, and advocates the need to develop inter-sectoral tools for sustainability analysis.

Experiences

The second half of the book examines a number of case studies, which explore some of the issues identified in the first half in more detail. Several of these case studies reflect the applied and regional traditions in geography and reinforce the overall significance that case studies may have in revealing some of the difficulties encountered in achieving sustainable tourism development (Hall and Jenkins 1995). Within the travel industry, which creates and sells tours to tourists, ecotourism has become the sector that is the most closely aligned with the concepts and goals of sustainable development. Several of the chapters in this section of the book discuss aspects of the ecotourism industry and the development of ecotour operations.

Chapter 8 by Lew examines trends in the Asia–Pacific ecotourism market. The chapter notes that this speciality travel industry has seen a broadening of its clientele and an increase in independent travellers (FITs) in recent years, with more people of varied ages and incomes taking ecotours, leading to a softening of the adventure element and a greater sensitivity to pricing. These new clients are also more aware of the environmental and social issues in the destinations they are visiting. Many problems and barriers to the development of ecotourism are identified in the chapter. One is a lack of knowledge about ecotourism in most of the travel industry and among most tourists, which is reflected in the political arena in a mistrust of ecotour operators by officials in some destinations. The ecotour industry itself lacks sufficient standards for quality and is experiencing pressure to expand into mass tourism. As more remote places become better connected to the outside world, prices are likely to fall, along with the loss of the natural environments and traditional cultures that form the basis of many ecotour experiences. The ecotour industry will therefore continue to evolve to reflect the changing economic and social conditions of both tourist generation countries and destination host countries.

Chapter 9 by Place looks at the value of ecotourism as a tool for rural development in Costa Rica. According to Place, Costa Rica has attempted to use its famous national park system as a magnet for nature-based tourism. Environmentalists and development professionals alike have viewed nature-based tourism, now formalised as ecotourism, as a win–win situation that encourages conservation of natural ecosystems while providing local populations with a sustainable economic base. Yet she argues that ecotourism embodies many of the contradictions inherent in capitalist development strategies and calls into question the sustainability of the enterprise, and also the meaning of 'development', when it systematically disadvantages those already at the bottom of the economic and social ladder – the rural poor.

Another case study of nature-based tourism, this time in a developed country, is provided by Forbes in Chapter 10. The chapter examines the Curry County sustainable nature-based tourism project in Oregon, in which tourism is being used as a response to problems of economic restructuring in the timber industry and the need to diversify local economies. In particular, the project focuses on issues of planning, product development, business training and marketing, and some of the positive and negative aspects of the project.

Chapter 11 by Charlton examines rural recreational public transport initiatives as a form of sustainable tourism in Britain. The chapter discusses the Devon and Cornwall Rail Partnership, based in southwest England, which has sought to develop and promote a series of local branch rail lines for both tourists and local residents. Significantly, this partnership has acted as a model for similar ventures elsewhere in Britain. According to Charlton, such partnership arrangements may be seen partly as a response to an uncertain and pressurised policy environment, especially on the part of local authorities. However, the same conditions generate short-term planning and funding for partnership ventures, thereby throwing some doubt on the very 'sustainability' of the tourism initiatives they have pioneered.

Chapter 12 by Hull examines ecotourism development on the Lower North Shore of Quebec, Canada. Residents on the Lower North Shore are responding to a fisheries crisis by identifying alternative economic development strategies, including a small-scale ecotourism industry centred around some of the oldest seabird sanctuaries in North America. Policy makers and communities hope that this type of tourism will generate revenue and employment while also protecting natural areas upon which local residents depend. The chapter examines the ability of ecotourism to act as a mechanism for sustainable economic development on the Lower North Shore of Quebec and analyses the potential role of ecotourism in contributing to seabird sanctuary management and to the local economy.

A growing body of literature treats ecotourism as a promising strategy for providing sustainable development. The main benefit of ecotourism is supposedly its potential for providing needed capital for local and national economies without exceeding ecological or cultural carrying capacities. Chapter 13 by Pobocik and Butalla examines the effects of independent and organised trekking tourism in the Annapurna Conservation Area in Nepal after ten years of management. Tourists, lodge owners and villagers in Nepal were surveyed to determine economic benefits, environmental consequences, and motivations of trekkers. Findings indicate that organised group treks provide a greater amount of income for the national government of Nepal and businesses in the regional centres of Kathmandu and Pokhara but offer little to local economies. Independent trekking is no more harmful to the environment than group trekking. However, it appears that trekking tourism is exacerbating fuel shortages and related environmental concerns despite the willingness by 80 percent of all tourists surveyed to share in the cost of correcting environmental problems and paying for alternative fuel supplies.

In 1991, New Zealand enacted the Resource Management Act (RMA), one of the world's first pieces of legislation that explicitly sought to enshrine the concept of sustainability in planning law. Chapter 14 by Page and Thorn discusses how the Act has affected the tourism industry and how it has been interpreted and implemented by local and regional government agencies. The chapter illustrates the difficulties that exist with implementation with reference to a survey of regional tourism organisations and local government bodies and concludes that despite the initial enthusiasm surrounding the RMA, substantial areas of uncertainty exist with respect to the longer-term acceptability of the Act and the resource consent process and its application to tourism.

It is somewhat of a paradox that although urban areas have been recognised as one of the 'most important type[s] of tourist destination[s]' (Law 1993: 1), they have to a large extent been excluded from discussions about sustainable tourism. This omission of urban areas from the discussion on sustainable tourism is difficult to explain but is at least partially due to the dynamic and complex nature of the city and the often fragmented nature of its tourism function. Whereas unsustainable development in pristine natural environments often results in dramatic failures, readily attributed to tourism, similar outcomes in an urban setting may go unnoticed, or at least the significance of tourism's role is seldom fully appreciated. Chapter 15 by Hinch aims to demonstrate the relevance of principles of sustainable tourism in the context of urban tourist attractions. Basic concepts and principles of sustainable urban tourism are reviewed in the context of the built, natural and cultural dimensions of the urban environment with specific reference to the case of Fort Edmonton Park in Edmonton, Alberta.

The final chapter, by the editors, draws together some of the themes and issues identified in the introductory chapter and highlighted throughout the book. This chapter therefore aims not only to indicate the substantial contribution that geography and geographers, and other social science disciplines, have made to the analysis of sustainable tourism development but also posits potential future connections that may be established between the various components of the discipline and what is a major, if not *the* major, issue in tourism studies today.

Chapter 2

Historical antecedents of sustainable development and ecotourism: new labels on old bottles?

C. Michael Hall

Although sustainable development has become something of a catch-cry of the 1990s, the concept has a long history. Indeed, the political conflict that has emerged over the way in which sustainability should be defined, managed and implemented is a reflection of a longstanding debate over the best use of natural resources in Western industrial society, which is, in turn, bound up within the wider framework of attitudes towards the environment. Geographers have been interested in the appropriate use of the physical environment by humankind since the middle of the nineteenth century and have also served to chart the history of environmental attitudes in Western and other societies. This research can offer profound insights into the manner in which exploitation of the environment occurs, the nature of environmental perceptions and conflict and behaviour relative to the environment, and the development and analysis of environmental policies.

Sustainability is an 'essentially contested concept' (Gallie 1955–56), that is, a concept the use and application of which is inherently a matter of dispute. The reason for this is the degree to which the concept is used to refer to a 'balance' or 'wise' use in the way in which natural resources are exploited. The appropriateness of such an approach and the very way in which 'wise use' is defined will depend on the values and ideologies of various stakeholders. However, the history of natural resource management over the last century would suggest that sustainable development is another term that has emerged in an attempt to reconcile conflicting value positions with regard to the environment and the perception that there is an environmental crisis that requires solution.

This chapter aims to offer a brief discussion of some of the early antecedents of sustainable development, particularly with respect to one of the earliest explicit linkages between tourism and the physical environment, the creation of national parks. In doing so, it seeks to emphasise that the political debate over sustainability is a continuation of a debate that has been occurring in industrial society since the 1870s between what might be described as economic conservationists and romantic conservationists. Furthermore, it also notes that the use of tourism to justify the setting aside of natural areas from other forms of economic development, such as agriculture, commercial for-estry or mining, also dates from this period. Such historical relationships are significant because they highlight not only continuity in the history of ideas but may also raise significant questions about the likelihood of success of present-day attempts to develop sustainable forms of natural resource develop-ment through such mechanisms as tourism.

Changing attitudes towards nature in Western society

Until the rise of the Romantic movement in the late 1700s, the Western ideal of nature was one in which there was an ordered and cultivated landscape in which wild nature was controlled and the boundaries of the wilderness made apparent. For example, the early European settlers in North America found themselves confronted by a harsh, forested environment reminiscent of that of Beowulf (Wright 1957). The forests were regarded as being a haven not for the settlers but for 'primitive' Indians, wild animals and beasts. Michael Wiggles-worth, a Puritan pioneer of New England, described the new world before him as

> A waste and howling wilderness,
> Where none inhabited
> But hellish fiends, and brutish men
> That Devils worshipped.
>
> Michael Wigglesworth, 1662 (in Adams 1966: 1)

The Puritan attitude towards the forest wilderness set the tone for North American attitudes towards wilderness for the next one hundred years and matched attitudes towards the environment by European settlers in the other new worlds of Australia and New Zealand. To the Puritans and other early settlers, the 'howling wilderness' was as much a state of mind as a state of fact. Although the discovery of the New World had initially raised expectations that an earthly paradise did exist, it was soon recognised that any anticipation of a second Eden was unrealistic. The American wilderness was not a paradise. The wilderness was something to be 'conquered', 'subdued' or 'vanquished'. If the settlers expected to enjoy an idyllic environment, then it would only be created through their own toil. It was only in its cultivated state that land acquired any value. The rural idyll was the desired landscape, a notion borne out in Thomas Jefferson's (1861) emphasis on the moral supremacy of the rural landscape as opposed to the moral degeneracy that could occur in either the towns or the wilderness.

The Romantic vision

The eighteenth century was an age of classification 'Insects, plants, animals and the races of man were divided into genera, species and sub-species. It was commonly supposed that this would lay bare the Divine Order or rational structure beneath the face of nature' (Honour 1981: 18). However, the result was 'entirely contrary'. The mechanistic and static conception of nature that characterised the Enlightenment was gradually replaced by one that was organic and dynamic. Romanticism emerged as the result of the decline of old orthodoxies. The physical embodiment of the Enlightenment in the form of the factory movement and mass production in which labour was objectivised and reduced to the status of a commodity was anathema to the Romantics. 'Romanticism was and is the antithesis of everything 'scientific' – logical behaviour, order, authority. . . . The romantics maintained that science was inadequate to explain all the phenomena with which man is confronted. They regarded these phenomena – understandable through intuitional, instinctive, and emotionally-based knowledge – as the most noble aspects of being human' (Pepper 1984: 77).

It would be grossly incorrect to characterise Romanticism as a comprehensive manifesto; rather Romanticism was an attitude towards life that valued the spiritual over the material. As Russell (1946: 653) noted, 'the romantic movement is characterised, as a whole, by the substitution of the aesthetic for utilitarian standards'. Nature, rather than being an object to organise and order, was upheld as a subject in its own right. As John Lorain (1825, in Glacken 1967: 44) wrote, 'the value of animal and vegetable matter is best seen in our lonely forests, where neither art nor ignorance has materially interfered, with the simple but wise economy of nature'.

In the intellectual climate created by the Romantic movement, wilderness and untamed nature lost much of their repugnance. 'It was not that wilderness was any less solitary, mysterious, and chaotic, but rather in the new intellectual context these qualities were coveted' (Nash 1967: 44). Landscapes such as mountains and wilderness, which once were landscapes of fear, now became landscapes of awe and admiration (Nicholson 1959; Tuan 1979; Honour 1981).

To the Romantics the New World was perceived as a new Eden in which man could draw close to wild nature. A cult of the primitive developed in which native peoples and the frontiersman, untouched by the civilised hand of European man, became archetypal romantic heroes. Contact with wilderness was believed to give man great strength and hardiness and an innate moral superiority over his more civilised counterparts (Fairchild 1928; Honour 1975). Moreover, in the writings of the Romantics, and more particularly the American transcendentalists such as Emerson and Thoreau, wild nature came to be endowed with a spiritual property, wholeness and wellness.

In Thoreau's interpretation of the natural world, humankind was a part of nature, not superior to it. Thoreau's Romantic ecology represented a revolutionary divergence from the bias against nature in some important strands of Western thought (Worster 1977). Thoreau was concerned with relation, interdependence and continuity, concepts which focused on the holistic structure of nature rather than the reductionist thinking of the day. In this new intellectual climate, wilderness became more and more important, the workings of wild nature rather than the works of man came to be seen as perfection. It was the notion of the perfection of wild nature that led Thoreau (1968: 11) to exclaim in 1851: 'Let me live where I will, on this side is the city, on that the wilderness, and ever I am leaving the city more and more, and withdrawing into the wilderness.'

The spiritual values of wilderness identified by Thoreau and the transcend-entalists led to the growth of demands to preserve the wilderness. To Thoreau, 'wildness and refinement were not fatal extremes but equally beneficent influ-ences Americans would do well to blend' (Nash 1967: 95). America's future, Thoreau believed, lay in the physical and metaphorical wilderness frontier of the west. 'The West of which I speak is but another name for the Wild; and what I have been preparing to say is, that in Wildness is the preservation of the World' (Thoreau 1968: 11–12) – a dictum that provided an inspiration for the legislative preservation of nature.

Preserving nature

The transcendentalists provided the intellectual legacy that laid the founda-tions for the preservation of wilderness and the establishment of a Romantic ecological vision of national parks and the environment that survives to the present day. However, one of the main influences in the development of an economic or 'progressive' basis for conservation and a direct legacy for the development of the concept of sustainable development was the publication of George Perkins Marsh's book *Man and Nature; or, Physical Geography as Modified by Human Action* in 1864 (Marsh 1965; see also Lowenthal 1958) (see Chapter 1).

Marsh's book contained two main theses: first, that when nature is left alone it is in harmony; second, that mankind impoverishes nature. In an alter-native interpretation of Genesis 1:28, Marsh (*ibid.*: 36) argued that 'the earth was given to him for usufruct alone, not for consumption, still less for proflig-ate waste'. The intention of Marsh was to demonstrate the need to balance man's use of the natural world. Influenced heavily by his observations in Europe, especially by the example of flooding caused by the clearing of forests in the Alps, Marsh identified major economic as well as romantic arguments for the preservation of nature.

To Marsh (1968: 14), 'man is everywhere a disturbing agent. Wherever he plants his foot, the harmonies of nature are turned to discords.' The attitude of Marsh to man's effects on nature was summed up in the quotation con-tained on the original title page of *Man and Nature*: 'Not all the winds, and storms, and earthquakes, and seas, and seasons of the world, have done so much to revolutionize the earth as MAN, the power of an endless life, has done since the day he came forth upon it, and received domain over it' (Marsh 1965: 1).

For Marsh, a scientific approach to the good husbanding and preservation of natural resources was essential for America's future. Marsh identified Amer-ica's long-term economic well-being as depending upon the maintenance of its renewable natural resources. The impact of Marsh's writings went well beyond America's shores. His central thesis of the need to restore or maintain the balance of nature that 'man' had disturbed, especially in the forest lands, was well publicised in Australia by leading newspapers such as the Melbourne

Age and the *Argus*. The *Argus* of 16 October 1865 noted that 'The conservation of the forest lands, and the extension and improvement of them, concern alike the landholder and the miner, and should occupy the attention of everyone who had leisure and means to become a co-worker with nature' (in Powell 1976: 62). However, recognition of the aesthetic or economic conservation dimensions of wilderness preservation were not of themselves sufficient to establish reserves. The first natural area national reservation to be set aside in the United States was the Arkansas Hot Springs in 1832. 'It was not scenically important, and was reserved to the government' because the springs 'were thought to be valuable in the treatment of certain ailments' (Ise 1961: 13). However, the reservation was not a 'park', nor was it a wilderness area, yet like so many of the national parks that would be established it did have tourism potential.

Tourism was the driving force behind the creation of the first national parks and conservation reserves. For example, Yosemite was ceded to the state of California on 30 June 1864 by President Lincoln as a state park 'for public use, resort, and recreation' (Nash 1963: 7). Tourism gave value to lands that were otherwise useless in terms of other forms of economic exploitation. For example, on introducing congressional legislation in 1864 to grant Mariposa Big Trees and Yosemite Valley to California, Senator John Connes of California assured his colleagues that the lands were 'for all public purposes worthless, but . . . [they] constitute, perhaps, some of the greatest wonders of the world' (in Runte 1977: 71). Similarly, in the designation of two million acres in northwestern Wyoming as Yellowstone National Park on 1 March 1872, Thomas Hayden, the leader of the geological expedition to the area whose report to Congress proved influential in the declaration of the park, noted the scientific significance of the area and the need for its preservation. However, he also 'assured them that Yellowstone seemed to be worthless for lumbering, mining, settlement or cattle raising' (in Runte 1972b: 6). It is notable that the House Committee's report emphasised Hayden's opinions on the economic value of the region and ended with the statement that a Yellowstone National Park would take 'nothing away from the value of the public domain' and 'was no pecuniary loss to the Government' (in *ibid.*: 6). Value instead came through tourism. It should therefore be no surprise that support for the establishment of a reserve was also found from the Northern Pacific Railroad Company (Sax 1976), which was routed close to the park.

The worthless or waste lands idea finds parallels throughout the Western world. The nucleus of present-day Tongariro National Park, New Zealand's first national park, was gifted to the national government in 1887, with the park being legally designated in 1894. The considerable delay between the deeding of the land of the Maori Chief, Te Heuheu Tukino, to the Crown and the actual establishment of the park reflected the government's concern that only 'worthless' land would be incorporated in the park. 'There had to be absolute certainty that land being added to the park had no economic value' (Harris 1974).

New Zealand is unique in that it was the first (and last) to reserve a national park in cooperation with its indigenous people. However, Te Heuheu's gift appears to have been made for political rather than environmental reasons. First, other tribes in the area were contesting the ownership of parts of the Tongariro area. Second, and perhaps more important, the rapid encroachment of the *pakeha* (colonial settlers) into the area had the potential to violate the extremely sacred nature of the three peaks to the Ngati Tuwharetoa tribe. Te Heuheu's son-in-law, Grace, a *pakeha* member of Parliament, had suggested several years earlier that Te Heuheu sell the region to the government, but had been refused. However, under increasing pressure from other tribes and *pakeha* settlers, Te Heuheu acceded to Grace's suggestion:

> Why not make them a tapu place of the Crown, a sacred place under the mana of the Queen? That is the only way in which to preserve them for ever as *places out of which no person can make money*. Why not give them to the Government as a reserve and park, to be the property of all the people of New Zealand, in memory of Heuheu and his tribe?
>
> (in Harris 1974: 53)

It is therefore somewhat ironic that, given the sentiments behind the park, the New Zealand politicians focused on the potential monetary value of the park. In discussing Tongariro National Park, the Minister for Lands, John McKenzie, was reported as telling Parliament that, 'anyone who had seen the portion of the country . . . which he might say was almost useless so far as grazing was concerned, would admit that it should be set apart as a national park for New Zealand' (New Zealand 1894: 579). In a similar fashion to the governments of Canada and the United States and the colonial governments of Queensland and Tasmania (Hall 1992a), the New Zealand government saw national parks as a means to develop areas through tourism, the aesthetic value of the regions being the attraction to the tourist. To quote the Hon. John Ballance on the Tongariro proposal: 'I think that this will be a great gift to the colony: I believe it will be a source of attraction to tourists from all parts of the world and that in time this will be one of the most famous parks in existence' (New Zealand 1887: 399).

The rise of progressive conservation

The year 1890 was notable not only for the creation of Yosemite National Park but also for an event that would have far greater impact on the popular consciousness of the United States – the closing of the frontier. The results of the census of 1890 indicated that for the first time centres of population stretched out across the continental United States. This did not mean that vast, empty spaces did not exist; rather, it indicated that America was becoming increasingly characterised by industrialisation and urbanisation rather than by the pioneer. For two and a half centuries, 'the frontier had been synonymous with the abundance, opportunity, and distinctiveness of the New World' (Nash 1968: 37). With the close of the frontier a form of cultural anxiety developed that focused on the need to retain links with the wilderness out of which the American nation had been created (Turner 1893, 1920; Mattson 1985).

Reaction to the loss of the frontier manifested itself in two ways: first, the rise of progressive conservation, in which the finite nature of America's natural resources was recognised (Hays 1957, 1959); second, the reinforcement of the perception of wilderness having spiritual values for the American people and the consequent rise of what Worster (1977) described as 'Romantic ecology'. The progressive conservation movement represented a 'wise use' approach to the management of natural resources, and its conservation motives were economic rather than aesthetic in intent. Hays (1959) saw three agencies as being the product of the movement: the Bureau of Reclamation, the National Park Service and the United States Forest Service.

Though the Forest Service was not founded until 1905, momentum for its creation had been building up in the two prior decades. Several bills relating to timber on public land had been introduced from the 1870s onwards, but in 1891 the president was given the power to set aside vast acreages in the public domain as forest reserves (Clarke and McCool 1985). Both preservationists and progressive conservationists saw the Forests Reserves Act of 1891 as a means of protecting wilderness areas. Preservationists led by John Muir wanted wilderness to contain no human activity that would be unsympathetic to the primitive nature of a wilderness area. However, progressive conservationists led by noted forester Gifford Pinchot and Theodore Roosevelt wanted forest lands to be managed on a sustained yield basis and were therefore in favour of timber harvesting, the building of dams for water supplies, and selective mining and grazing, all in the name of conservation. In a statement that recalls much of current debates over sustainability, Gifford Pinchot (1968: 9) stated in 1910 that

> The first great fact about conservation is that it stands for development. There has been a fundamental misconception that conservation means nothing but the husbanding of resources for future generations. There could be no more serious mistake. Conservation does mean provision for the future, but it means also and first of all the recognition of the right of the present generation to the fullest necessary use of all the resources with which this country is so abundantly blessed. Conservation demands the welfare of the country first, and afterward the welfare of the generations to follow.

Initially, there was a reasonable degree of correspondence in the views of Romantic ecologists, such as John Muir, and economic conservationists, such as Pinchot. Muir, for instance, wrote in 1895 that 'it is impossible in the nature of things to stop at preservation. The forests must be, and will be, not only preserved, but used, and . . . like perennial fountains . . . be made to yield a sure harvest of timber, while at the same time all their far-reaching [aesthetic and spiritual] uses may be maintained unimpaired' (in Nash 1967: 134–135). However, over time a split occurred between the various parties as to how conservation reserves should be managed.

Pinchot and the progressive conservationists advocated the 'wise' use of natural resources, while the preservationists continued to focus on the aesthetic and spiritual qualities of forest wilderness. As Fernow (1896 in Nash 1967: 137) wrote in *The Forester*, 'the main service, the principal object of the forest has nothing to do with beauty or pleasure. It is not, except incidentally, an object of esthetics, but an object of economics.' Such a viewpoint was anathema to the preservationists. Muir believed that 'government protection should be thrown around every wild grove and forest on the mountains' in order to preserve the 'higher' uses of wilderness (in Nash 1963: 9). The problem that faced Muir, which exists to this day, is that the existence of 'undisturbed' wild nature is incompatible with productive forest management.

The creation of the United States Forest Service in 1905, with Pinchot at its head, marked the institutionalisation of progressive conservation in the United States government (Richardson 1962). Government forestry, and wider involvement in the management of natural resources, in America was founded upon Pinchot's vision of academic forestry – 'that is, the scientific management of the timber resource according to the principles of wise use and sustained yield' (in Clarke and McCool 1985: 36). These principles have strongly influenced not only forestry practices throughout the world but the wider field of resource management, including tourism's use of the environment.

Primitive ecotourism: the national parks and tourism

The aim of the National Park Service was to include in the park system all areas that contained 'scenery of supreme and distinctive quality or some natural feature so extraordinary or unique as to be of general interest and importance' (in Buck 1921: 52). However, the mandate of the National Park Service provided a paradox that lingers to the present day: the service was meant to provide enjoyment for the people and hence attract them to the parks, while simultaneously it was supposed to keep the parklands in an unimpaired state. This dilemma has been repeated in many park management bodies around the world. Because of the low visitor levels, such a situation was not harmful to the parks in their early days. Indeed, in an effort to promote the national park cause the service eagerly took up a 'see America first' campaign.

The value of the national parks to tourism was stressed by the first director of the National Park Service, Steven Mather, and his assistant, Horace Albright (Shankland 1970; Swain 1970; Sax 1976). Appealing to the utilitarian spirit, Mather often invoked the profit motive in relation to the national parks. In 1915, Mather (in Runte 1979: 103) claimed that 'our national parks are practically lying fallow, and only await proper development to bring them into their own. . . . A hundred thousand people used the national parks last year. A million Americans should play in them every summer.'

Mather and Albright actively encouraged automobile users to visit the national parks by extending and upgrading park roads and supporting the upgrading of highways. The railways also continued their strong support of the parks. Lois Hill (in Foresta 1984: 24) of the Great Northern Railroad noted that 'every passenger to the national parks represents practically a net earning.' It is therefore not surprising that seventeen western railroad companies contributed to the publication and distribution of a glossy publicity portfolio of the national parks in order to promote tourism activity (Buck 1921). However, it is somewhat ironic that the increased popularity of automobile transport to the parks led to the decline and eventual failure of many park railroads (Runte 1974).

Mather was 'no primitive who wanted to curb mass use', but neither did he want Coney Island-type amusement parks established in the national parks; rather, appropriate tourist facilities were regarded as enhancing the appeal of the parks (Foresta 1984). Mather's attitude to the national parks is probably best summed up in Secretary Lane's letter of 13 May 1918, in which the administrative policy for the parks was outlined:

> First . . . national parks must be maintained in absolutely unimpaired form for the use of future generations as well as those of our own time; second . . . they are set aside for the use, observation, health, and pleasure of the people; and third . . . the national interest must dictate all decisions affecting public or private enterprise in the parks.
>
> (Secretary Lane, letter to Steven Mather 1918, in Ise 1961: 195)

The principles by which the national parks were managed established the wilderness idea within the national parks – the notion that the parks 'must be maintained in an absolutely unimpaired form'. However, notions of strict wilderness preservation were in many ways at odds with the desire to attract tourism and recreation interests to the parks. Mather and Albright attempted to travel a middle path, 'usually they would allow nuclei of intensive visitor services in the parks, make those nuclei and some of the most spectacular sites . . . accessible by high-grade roads, and leave the rest of the parkland – most of it – as wilderness' (Foresta 1984: 20).

By the late 1920s, concern for endangered species led the supporters of national parks to recognise that parks contained more than just scenery, and that protection of plant and animal associations was an integral part of the nature of environmental conservation. The natural environment was now evaluated in what Runte (1979: 106) described as 'complete conservation', in which natural areas were recognised as being able to support a wide range of values: recreation, spiritual renewal, religious experience, health and ecology, a situation synonymous with present-day perceptions of the values of wilderness (Hall 1992a).

The first move towards an ecologically based national park was the creation of the Everglades National Park in Florida in 1934. Runte (1979: 108–109) claimed that 'For the first time a major national park would lack great mountains, deep canyons, and tumbling waterfalls, preservationists accepted the protection of its native plants and animals alone as justification for Everglades National Park.' Runte is correct to note that the park was substantially different from existing national parks in terms of its landscape and ecological characteristics. Yet he is incorrect to assume that its ecology alone was responsible for its preservation. The scenic qualities of the region, so important for tourism, were still an important force behind its creation. As the fact-finding committee of the National Parks Association noted:

> even granting the . . . limitations as to the scenery of parts of the region, there are extensive areas where even the most casual observer can hardly fail to be gripped and inspired by a sense of power and vastness of nature, essentially akin to the feelings inspired by great scenes in our existing National Parks yet arising out of elements so different from these – indeed so wholly unfamiliar to the experience of most visitors to the National Parks – as to have the special force of novelty.
>
> (Olmsted and Wharton 1932: 143)

Nature alone was not yet reason enough for the protection of a natural area.

Antecedents

A review of the historical antecedents in the sustainable development of natural resources generates a number of significant insights into the present-day issues that surround sustainability. First, debate over the sustainable development of natural resources in industrialised countries dates from the middle of the nineteenth century and cannot be seen as a new policy issue, at least at the local or national level. Second, tourism has long been a key factor in the just-ification for environmental conservation. Third, there has been no easy middle path in attempting to find a balanced use of natural resources. Political reality, rather than ecological reality, has been the order of the day.

Present-day discussions on the nature of sustainable development have their antecedents in the debates that have taken place for well over a century, and have been most clearly manifested with respect to tourism in the establishment of national parks. The history of national parks, which contained the apparent paradox between visitation and conservation almost from the outset, contains many lessons for the use of ecotourism as a conservation and development tool in the 1990s: most particularly, the problem of controlling numbers once people have been attracted and when, often, the political rationale for the park is primarily based on visitation. As Hall and Wouters (1994) observed with respect to tourism in the sub-Antarctic islands, from an ecological perspective, sustainable tourism means conserving the productive basis of the physical envir-onment by preserving the integrity of the biota and ecological processes and producing tourism commodities without degrading other values. Having no form of tourism in the sub-Antarctic islands may well be the most advisable management strategy in terms of the ecological integrity of the islands. How-ever, it is also unrealistic. In order to ensure that natural areas are preserved we must, somewhat paradoxically, allow people to visit these wild places so that policy makers can be persuaded to maintain their reserve status (Booth 1990). Vicarious appreciation through books and documentaries is important, but it is not sufficient to create a groundswell of public opinion for preservation (Hall 1992a).

Undoubtedly, the perception of crisis because of such environmental issues as DDT, oil pollution, global climate change and deforestation, which has affected popular opinion on the environment, has played a major part in raising the debate over the conservation and use of natural resources. Tourism was, and still is through the present focus on ecotourism, seen as a mechanism both to conserve the environment and develop peripheral areas, even though the effectiveness of management tools is still highly debatable (see Chapter 7). Furthermore, unlike the nineteenth and early twentieth centuries, the environment is now a global issue that requires both an international response and a global analysis (see Chapters 4 and 5). Finally, unlike the earliest attempts at natural resource conservation, there is also a growing recognition that the conservation of any landscape is culturally driven and recognition must be given to the cultural values of landscape, particularly those of indigenous peoples (see Chapter 6). Nevertheless, the core issues that surround sustainable development, particularly with respect to tourism, still remain; that is, a 'balanced' form of development that allows us to conserve the natural environment while also allowing it to be exploited so as to ensure economic growth.

Undoubtedly, one of the major drivers behind the manner in which natural resources are consumed, including through tourism, and managed is the contemporary global capitalist system and the values that underly it. Yet, while the overthrow of this system would be an 'ideal' solution for some in their desire to provide a solution to the undoubted ecological crisis that the world is facing, it is also unrealistic. History has indicated that such radical change will be unlikely, at least in a global setting, for the foreseeable future.

Economic or progressive conservation was the dominant metaphor for natural resource management, including tourism, until the 1930s. For example, in 1915, the Canadian Commission of Conservation suggested that 'each generation had the right to profit from the interest on nature's capital, but that this capital had to be maintained intact for future generations to use in a similar fashion' (Vaillancourt 1995: 222). Similarly, in 1948, the International Union for the Conservation of Nature and Natural Resources (IUCN) was founded on the premise that both nature and its resources should be protected for the benefit of existing and future generations. For much of this century, and particularly during the 1960s, the idea of sustained yield was paramount in natural resource management. At the 1972 United Nations Conference on Human Environment, the idea of 'eco-development' was put forward. In the 1970s and early 1980s, the idea of coordinated and integrated development and resource management was put forward, with a corresponding incorporation into the tourism lexicon. In the mid-1980s, the ecological principles of community-based development were propounded. Now we have supposed 'alternatives' in the form of sustainable tourism development and ecotourism. Undoubtedly, the ideals and principles behind sustainable tourism development and ecotourism, which, at their most crude, translate to not shitting in one's own nest, and, hopefully, on a global scale, not shitting in anyone else's, are important principles to aspire to. Yet we all have to shit. Therefore, sustainable tourism development and ecotourism are yet another restating of an old problem. The ideals they represent are undoubtedly important. However, the political reality that surrounds tourism and development is such that ideal solutions are not implemented on a scale beyond the exceedingly local. The lesson of history is that the path of sustainable tourism development will be one of muddling through. Unless the scale of analysis and action can be turned to the bigger picture in both space and time, it is unlikely that tourism will ever become truly sustainable beyond anything more than the most local of cases.

Chapter 3

Sustainable tourism – looking backwards in order to progress?

Richard Butler

Introduction – old wine in new bottles?

It is now more than a decade since the term sustainable development made its way into the common lexicon of politicians, planners, developers and the public at large. The widespread acceptance of the term is both satisfying and disturbing to those seriously interested in the long-term viability of the physical and social environments in which mankind lives and operates. One of the most frustrating, but not unexpected, aspects of this adoption of the *concept* of sustainable development is the fact that the *implementation* of the idea has been much less successful. As will be noted below, and as has been discussed by other authors in this volume and elsewhere (Clarke 1997; Craik 1995; Nelson, Butler and Wall 1993; Stabler 1997), this can be explained at least in part by problems in uncertainty and ambiguity over the meaning of the term.

A second fascinating aspect of the popularisation of the concept is that to many of its supporters it appears to be regarded as new, when in reality the fundamental principles are well established and have been common to many societies for a long time. One can argue strongly that these principles are close to those of the original idea of conservation as it emerged at the end of the nineteenth century, espoused so well by Gifford Pinchot and his school in the first decades of this century (Nelson 1973) (see Chapter 2). More recently, similar principles were espoused by Dennis Meadows and his colleagues in their report (Meadows *et al.* 1972) to the Club of Rome more than a quarter of a century ago, amid the first post-war flush of environmental concern following the publication of Rachel Carson's consciousness-raising volume *Silent Spring* (Carson 1962). Several years later, Garret Hardin published his controversial essay on *The Tragedy of the Commons* in *Science* (Hardin 1969), which suggested the inevitability of environmental decline in the absence of assigned responsibility for resource protection (a theme taken up in the context of tourism by other writers much later such as Butler (1991) and Healy (1994)). The same year that the report by Meadows and his team was published, over one hundred nations met in Stockholm for the first global conference on the environment, with clear antecedents to the 'Rio summit' two decades later.

All of these schools of thought raised one common problem, namely, that current generations were imposing too great a demand upon the natural environment to allow it to continue to reproduce and maintain itself at its previous level of stability. In other words, the primary concern was with the carrying capacity of the environment, normally in terms of the physical environment, and what was seen as overuse of that environment. The *World Conservation Strategy* (IUCN 1980) laid the basis for the fundamental message of the *Brundtland Report* (WCED 1987), namely, that development and conservation were both necessary principles on which the future pattern of human activity should operate and should be planned together in an integrated manner that reflected ecological and human processes and requirements. The call for sustainable development contained two additional elements that had not been explicit in the earlier discussions on wise use of resources, namely ethics and equity. These have received less attention and public discussion than other aspects of the concept, and their relevance to tourism in particular is discussed later.

Just why this call for action and acceptance of change by the WCED in 1987 should have received such a positive response is unclear (Muller 1997). One may suggest, perhaps somewhat cynically, that sufficient time has passed since the public concern of the 1960s for the current generation to see such ideas as new, rather than a repeat of a call that has been made periodically throughout mankind's presence on the Earth, from Marcus Aurelius through to Thomas Malthus. Present-day society seems enthralled with identifying what are regarded as new concepts or ideas and imbuing them with overwhelming attributes and abilities to correct society's problems. Few concepts have received the overwhelming endorsement given to sustainable development, however. Since its introduction in 1987, it has received enthusiastic support from more than a hundred governments and innumerable non-governmental agencies and other groups, as well as academic endorsement and apparent general public sympathy, at least in developed world countries (Kirkby *et al.* 1995). This author still has some doubts as to just how strong that support actually is, even in Western European countries, among the general public. Overall sympathy for the goals of the concept does not necessarily translate into acceptance of costs and sacrifices that actual application may entail. In some traditional tourist destination countries, while sustainable development and sustainable tourism may be officially promoted, rarely is more than lip service paid to the application of the concept, especially where this would involve reducing the numbers of tourists, or more importantly the tourist spend and employment generated (Butler 1991).

The overwhelming appeal of sustainable concepts lies in the generality of the concept and the fact that the true costs of the implementation of the concept have never been spelt out. Where the costs are perceived to be a reduction in development, and in tourism terms, fewer tourists, less employment and reduced income, then the concept is not supported enthusiastically, or is interpreted in terms of purely economic sustainability, which means that the primary concern is with maintaining the long-term viability of the economy of the region being considered, rather than the viability of the physical and social environments. In areas that currently experience low standards of living, extremely low incomes, overpopulation and resource scarcity, such mundane concerns as survival perhaps deserve more consideration than they have had to date in the rush to impose the sustainable doctrine by an overly moralistic developed world. Thus political and economic summits in locations such as Rio de Janeiro, in close proximity to some of the worlds' worst environmental and social problems such as the clearing of the Amazonian rain forest and urban slums, where leaders of countries and organisations discuss sustainability in settings that are as far from sustainable as is imaginable, do little to improve the acceptance of change and sacrifice at local levels that would be necessary to achieve global sustainable development. Despite this, international organisations continue to call for the implementation of sustainable principles in tourism planning and produce guidelines and directions for such implementation (McIntyre 1973; WTO 1997c).

Sustainable development and tourism

The adoption of the principles of sustainable development to tourism has been rapid and widespread, although implementation of the practice has been much more limited. Many conferences have been held on this theme, and a vast and rapidly increasing number of publications on the topic now exist (see, for example, Globe 90 (1990)). The World Conference on Sustainable Tourism, held in Lanzarote, was only one of a large number that produced recommendations on the application of sustainable development principles to tourism, in that case a 'Charter on Principles and Objectives for Sustainable Tourism' (Martin 1995 in France 1997: 13). The term 'sustainable tourism', rightly or wrongly (Butler 1993b), has become widely accepted as meaning tourism that is developed and operated in such a manner as to follow these principles. It has been sold at various levels as being appropriate and morally correct as well as being environmentally suitable, and thus has high appeal to tourists and to decision makers in the tourism industry, in both the public and private sectors (Wheeller 1993).

Where it has been adopted in the tourism industry, it has tended to be accepted for three reasons: economics, public relations and marketing. On the one hand, some aspects of sustainability as applied to tourism can be cost-effective and, in fact, reduce costs. Encouraging guests to conserve water, power and labour by not requesting that all of their towels be washed each day not only saves the accommodation unit money, but also allows it to achieve the second goal, good public relations. To excuse the accommodation unit from washing all towels each day allows a guest to feel good by making a positive contribution to world environmental well-being, and also suggests that the accommodation unit is serious about supporting sustainable principles. This can only be applauded in terms of the overall effect; any reduction in environmental impacts is laudable, even if the motives are slightly less noble. However, it does nothing to help correct a major problem, if for example, the accommodation unit has been built in an inappropriate location, or if the major concerns and impacts relate to the tourists' activities outside the accommodation, as is often the case (Butler 1993a).

It is of course in the marketing of sustainable tourism that the tourism industry has achieved remarkable success through its efforts in promoting the concept. One cannot avoid the feeling that this is due almost entirely, at least initially, to the correct perception that sustainable development is viewed as 'a good thing' by the media, and hence the public, and that sustainable products would be enthusiastically purchased. Thus began what Wheeller (1993) has so aptly called 'ego tourism'. Almost any non-mass form of tourism, and perhaps even some forms of mass tourism, gain in acceptability by being labelled 'sustainable', and in extreme cases some tourist developments are now claimed to be sustainable before they have even opened. As noted below, the final verdict on the sustainability of an operation cannot be made for many years, but verifiability has rarely stopped the marketing of a concept and associated developments. What is normally meant in such cases is a less environmentally impacting form of development than might otherwise have been the case. While the development in question may be a significant improvement over what existed before, or what might have been developed given past experience, the current state of knowledge about impacts, and the reality of impact assessment in most countries, means that we rarely know what the impacts of existing developments are going to be, let alone knowing what the impacts of developments not completed would have been (Butler 1993a; Ding and Pigram 1995; Goodall 1992). Thus, while some developments may have moved significantly towards sustainability, to claim that they are sustainable is clearly at best premature, and possibly completely inaccurate.

Sustainable tourism – a whole or a part?

The concept of sustainable development as discussed in the *Brundtland Report* (WCED 1987) is a holistic one, and this element is at the heart of successful adoption of the concept. The traditional separation of development and conservation (and preservation) has inevitably led to division and disagreement between proponents of these two approaches, and this problem was discussed at length in the report. It is depressing, therefore, to witness the apparent abandonment of the necessity of maintaining an integrated and comprehensive approach to global progress by maintaining traditional sectoral approaches (e.g. Priestley, Edwards and Coccossis 1996; France 1997). To talk of sustainable tourism in the sense that tourism could (and should) achieve sustainable development independently of other activities and processes (Croall 1995) is philosophically against the true nature of the concept, as well as being unrealistic.

It is clearly illogical and unrealistic to contemplate sustainability in any one sense alone, such as economic sustainability, and equally, one might argue that it is inappropriate to discuss sustainable tourism any more than one might discuss any other single activity. Given the fact that at the global scale we are dealing with a closed system, clearly we cannot hope to achieve sustainability in one sector alone, when each is linked to and dependent upon the others.

Thus calls for sustainable tourism to be developed irrespective of whether other, interrelated, segments are to be sustainable or not is inappropriate and contradictory. Despite this apparent inconsistency, various authors have identified different forms of sustainability in the context of tourism. Coccossis (1995) suggests that there are at least four ways in which to interpret tourism with respect to the principles of sustainable development. He describes these as relating to economic sustainability, to ecological sustainability, to the long-term viability of tourism, and to accepting tourism as a part of an overall strategy for sustainable development. Others have discussed different elements of sustainability in the context of tourism (Pigram 1990; Craik 1995; Bramwell *et al.* 1996). Bramwell *et al.* (1996: 5) identify seven dimensions of sustainability when reviewing the principles and practices of sustainable tourism management. These are environmental, cultural, political, economic, social, managerial and governmental, and it is not unreasonable to assume that many of the researchers, decision makers and others involved in each of these dimensions are likely to have their own different interpretations of the concept. Bramwell (1996) is one of the few authors who acknowledge that sustainable development, and in the context in which he is discussing it, sustainable tourism, cannot be separated from the value systems of those involved and the societies in which they exist. He correctly points out that this failure to appreciate the value-laden nature of the concept has helped to perpetuate the misunderstanding that sustainable development is a well-understood concept with a common meaning to all proponents.

Although the authors of *Our Common Future* (WCED 1987) wrote of the human environment as well as the natural environment, the initial focus on sustainable development was on the physical environment and its capacity to absorb the demands made on it by various forms of economic activity, including tourism. This focus on the physical environment has caused concern with respect to the implied or specific neglect of social and cultural issues, and these concerns have been well articulated by several authors, ranging from Walter (1982) over a decade ago to Craik (1995) more recently. Other recent publications have also drawn attention to the need to include all environments in any discussion of sustainability. It should be clearly recognised that if sustainable development, in the sense suggested by the Brundtland Commission (WCED 1987), is to be achieved, it can be achieved only if all environments and elements are dealt with simultaneously and from an integrated and holistic standpoint and not on a sectoral basis.

In many ways, the focus on the physical environment is easy to comprehend and it mirrors the process of impact assessment, with which it has much in common. Early studies on impact assessment in the 1960s dealt almost entirely with the physical environment, and the process of identifying and predicting the effects of development became known as environmental impact assessment. Only after a decade or more did the process become broadened to include social and cultural effects of development and the name changed to impact assessment. This is not meant to imply that economic impacts were not studied; on the contrary, economic impacts are normally the driving force behind development, certainly development by or in the private sector, which includes most tourism developments. Economic impacts are normally studied prior to development being undertaken, although the findings of such studies are often proprietary and rarely make the academic press. Similarly, in the field of tourism, studies of the economic impact of tourism are almost as old as those of tourism itself, while attention to the environmental and social impacts of tourism did not really begin until the 1970s (Butler 1989). Thus it is not surprising that the first calls for sustainability did not explicitly recognise the need to consider economic, social and cultural environments, or the fact that the same principles apply there as they do to the physical environment.

Sustainable tourism – the problem of scale

The application of the principles of sustainable development to tourism as a distinct sector inevitably raises difficulties. While the original call for sustainable development (WCED 1987) discussed the application of the principles with respect to a number of contexts, tourism was not specifically discussed. Thus a generalised statement of principles relating to the global environment has been adapted to a specific but unmentioned sector of that environment. It is not surprising, therefore, that there has been confusion and disagreement over what the principles really are in the context of tourism, and how they may be put into practice (Wheeller 1993; Clarke 1997; Stabler 1997). It can be argued that to apply the principles to any single sector is unrealistic and that sustainability in that sense is unachievable. Since the global environment represents the only complete discrete system, it is really only at that scale, if any, that true sustainability can be achieved.

The issue of scale is one that has been rarely discussed in the debate over sustainable tourism, and yet it is fundamental to the successful application of the principles. Just as the dimensions of the effects of tourism can only be assessed and managed in the context of a specific defined region, so too can sustainability only be understood and defined relative to a destination. The complete set of impacts of any activity can never be identified except at the global scale because of the issue of leakage. Only at the global scale is there no leakage or impact outside the area being considered. As sustainability is inextricably linked to the effects of the development being considered, in this case tourism, then the same general principles that apply to impact assessment can be argued to apply to sustainability. Thus, just as we can never measure the complete set of impacts of any development at anything less than the global scale, neither can we hope to achieve complete sustainability at anything less than this scale.

The most easily demonstrated example of this is in the context of travel involved with tourism. While the undesired impacts of tourism at a specific location may be reduced to such a level as to make them much less harmful to the human and physical environments in the context of the destination under consideration, the full impacts of those tourism activities occur not only in the destination but in other areas also, extending as far as the region of origin of the participants. Thus, for example, while the undesirable effects of tourism at a destination may be greatly reduced and mitigated, the impacts caused by travel to the destination, for example, may not be so diminished. In the overall context, therefore, such a development could not be regarded as being sustainable. The fact that unsustainable impacts from such a development occur in other regions than the destination tends to be ignored in the rhetoric claiming sustainability for a specific development. Arguing that tourism in a specific location is sustainable is frequently misleading and optimistic at best. Tourism in a specific area may have moved some way towards being sustainable and in reality that may be the best that can be achieved.

The timescale involved

It is perhaps in the case of the timescale involved that sustainable development offers the greatest difficulties in the context of tourism. The focus of sustainable development principles on future generations and long-term viability means that inevitably any form of development can only be judged sustainable or unsustainable after a long period of operation, when it can be ascertained if the demands of the activity have not prejudiced the needs of what were future generations when the development began. Thus claims that specific forms of tourism are inherently sustainable or that specific developments have achieved sustainability are at best premature and in many cases not only misleading but often wildly optimistic.

The crux of this discussion is that the application of sustainable development principles to tourism, while appropriate and praiseworthy in principle, has, in reality, given rise to unrealistic expectations and demands. Tourism, like any single sector of the economy, or any single region, cannot achieve sustainability in the sense in which the term has become defined, and so sustainable tourism becomes a misnomer. The commonly accepted definition of sustainable development, which is based on the *Brundtland Report* (WCED 1987), has not been translated effectively into action. There is, therefore, a clear lack of consensus about the way in which this definition should be translated into the management of people, resources and environments in a manner that would achieve universal acceptance. In many cases, what has been called sustainable development is at best wishful thinking, and one has to agree with a report on Scottish tourism, which honestly, if regretfully, concluded that no example of the successful application of sustainable development of tourism had been found (STB 1994).

Second, as already argued above, a single sector or single region cannot exist divorced from other sectors and environments, and thus true sustainable development can be achieved only at the global scale. Third, we have very little idea of the needs or even the true preferences of current generations for tourism, and no reliable idea of the needs of future generations, on which sustainable development is supposed to be based. Tourism forecasting is normally based on current levels being projected into the future, adjusted in line with forecasts of changes in the independent variables that affect tourism (Smith 1989). Tourism and tourism forecasts have never been based on needs, either of current or future generations, and as the majority of the world's population does not participate in tourism, in most cases because of lack of resources and time, it is somewhat inappropriate to talk of tourism as a 'need' – an important element in the global economic and social picture certainly, but hardly a need in the context, for example, of Maslow's oft-quoted hierarchy (Pearce 1993).

Fourth, even if we were capable of producing accurate numbers by which to express the needs of potential future tourists, society at large has never tackled the problem of prioritising tourism versus other sectors at the world and national levels. There are key questions to be answered in this context, including how we could balance the 'needs' of existing and potential tourists against the needs of local populations for resources and space, who should be charged with formulating such an equation (those in origin areas, present and future, or those in destination areas, present and future), and what we should do if, as almost inevitably will occur, there is disagreement over the results.

This last issue is directly related to the two elements that make sustainable development somewhat different from earlier calls for changes in attitudes towards development and the environment, ethics and equity (Lemons and Brown 1995). Discussion of the ethics of sustainable development has been muted and rather replaced by an apparently general tacit acceptance that the concept represents a kinder and more moral or ethical approach to development. There has been little challenge of the moral appropriateness of sustainable development, or sustainable tourism in this specific context, although the difference between the concept and the reality has been most effectively challenged (Wheeller 1993; Stabler 1997).

The issue of whose ethics should be used in formulating policy has received little discussion in the context of tourism. As in so many cases where a concept emerges from Western societies' recognition of the errors and effects of their ways, particularly with respect to the environment and the less developed world, it is Western values and ethics that provide the base for the implementation of the concept. In part, some of the opposition to such implementation has arisen because of reaction to the perceived further imposition of Western values and ethics on other societies, particularly when the benefits from the causes of the problems have already accrued to the same Western societies. In the case of tourism, what is called sustainable tourism is sometimes regarded as synonymous with 'alternative tourism', 'appropriate tourism', 'sympathetic tourism' or even 'ethical tourism' (Smith and Eadington 1991), awarding sustainable tourism a higher moral ground than other forms of tourism, particularly 'mass' or 'conventional' tourism. Such a situation should be a cause for concern, since it carries with it the implication that to be opposed to sustainable tourism is to be unethical and to support inappropriate forms of tourism (Wheeller 1993). In reality, 'mass' tourism may be much more appropriate and less harmful in many respects to both physical and human environments, and much more beneficial with respect to the economic environment in specific situations than supposedly sustainable or 'new' (Poon 1993) forms of tourism. Such discussions have appeared only rarely in the literature to date.

The other issue, that of equity, has a much longer history in resources management, and by implication in discussions on the management of tourism and recreation resources (Pigram 1983). Most calls for conservation and indeed preservation of resources and environments have had as their base a concern for the long-term survival of those resources and environments, and such principles are central to the policies of many national park agencies (Nelson and Butler 1975). The Canadian National Parks Policy, for example, declares that the parks are protected 'for the education, enjoyment and benefit of present and future generations' (Parks Canada 1994: iii). In the *Brundtland Report* (WCED 1987), the issue of inter-generational equity was made explicit as shown in the definition of sustainable development quoted earlier, and in the discussion on sustainable tourism the needs of future generations have sensibly included the needs of future generations of local residents of destination areas as well as, or often above, the needs of future generations of tourists. As noted earlier, however, there has been very little research on needs in the context of tourism and even less on the future needs of future generations of local residents. The proponents of tourism development have only begun to take into account the preferences and needs of current residents of destination areas relatively recently. As discussed earlier, the relative priority of the needs of present and future generations of residents to the needs of present and future generations of tourists has not been addressed to any significant degree, nor has it been settled who should decide such priority or on what grounds. Thus equity in the context of sustainable tourism has not been satisfactorily resolved in theory and is rarely discussed in practice.

Conclusions – looking backwards rather than forwards?

If it is to achieve a measure of sustainability that is a significant improvement over its general current position, then tourism must attack the major existing problems of sustainability in the industry rather than concentrating on minor new developments. These major problems, in terms of magnitude and global seriousness, stem from mass tourism, if for no other reason than the sheer volume of numbers involved and hence environmental and social impacts that result. They represent, for the most part, the results of poor planning and development of tourism by current and previous generations of decision makers and developers, although, it should be said, not all of such developments were without support and encouragement from local residents of the areas affected at the time of development.

The personal hygiene problems referred to in the previous chapter represent good examples of this. Sewage and attendant pollution problems are frequently significant manifestations of the effects of tourism and other forms of development on destinations. One needs to question and evaluate which would be best for a destination environment and its human population: development funds being made available to resolve a sewage problem caused by mass tourism at a destination; or supporting several small-scale ecologically and socially sympathetic and appropriate tourist developments in other areas. In most cases, solving the sewage problem in the mass tourism development can be argued to be much more beneficial in both local and global terms. It would represent, in this example, a current problem that has environmental and social, as well as potentially economic, implications for the destination (tourists may not come if an epidemic occurs as a result of polluted water). Its solution could prove beneficial to the permanent local population by solving the problem of their sewage also.

The development of supposedly sustainable small-scale tourism facilities, on the other hand, would presumably attract additional tourists, do nothing about the existing sewage problem in the mass tourism destination, and would, inevitably, lead to other tourism impacts in the newly developed locations. These impacts might well be less than they would have been had conventional mass tourism been developed in these locations, but that is hardly an acceptable answer. In most cases, mass tourism would not have been developed in those locations. A more serious question, not often discussed, is whether tourism should be developed in these locations at all. As the previous chapter has noted, in some areas, the best strategy for the environment, the local population and future generations may be no tourism development at all, rather than further development, even if it is supposedly sustainable. Thus praising small-scale additions to the tourist scene (Croall 1995) while ignoring more significant ongoing problems does little to persuade the critical observer that many players are really serious about the application of sustainable tourism principles at a level that will make a great deal of difference.

If we are to move any way towards achieving the goals of sustainable development in tourism, how ever ill-defined those goals may be in practical terms, little progress will be achieved at either local or global scales by concentrating on future developments. Major improvements in environmental and social (and probably economic) conditions in existing destinations will only be achieved by tackling the current and longstanding problems of earlier developments, in other words, by looking backwards. The causes of the problems lie in the past, even though the solutions and improvements lie in the future, and the problems themselves will only increase in magnitude and number if we ignore them in favour of more attractive politically correct and supposedly sustainable developments. We cannot be selective about sustainable development, thinking globally and acting locally will work only if local actions are part of an integrated holistic approach and include solutions to past problems. They must not ignore the fact that tourism is part of the global system and cannot be tackled in isolation, spatially, economically or temporally. In the case of sustainable tourism, the past holds the key to the future.

Chapter 4

Tourism and sustainable development: exploring the global–local nexus

Simon S. Milne

Much of the existing literature on tourism and development (sustainable or otherwise) is highly descriptive in nature and has been weakened by a continued fascination with theoretical models that either predict linear paths of destination evolution or speak of radical paradigmatic shifts from an 'old' to a 'new' tourism. This chapter argues that researchers need to develop a more rigorous and responsive theoretical framework if they are to understand better tourism's ability to generate sustainable development (SD). It also contains some thoughts on what sort of research agenda may lead us to a better understanding of the connections between SD and tourism and emphasises human agency, stakeholder relationships and broader theoretical constructs at the expense of detailed discussions of environmental impacts and methodological developments.

The chapter begins by outlining what is meant by SD and how this rather 'slippery' concept has been applied to tourism studies. It argues that while truly sustainable tourism can probably never be achieved, it remains an ideal that we must strive to attain. A tentative theoretical framework through which to view tourism's interaction with the development process is then presented – with a focus on the global–local nexus that moulds the industry's relationship with people's everyday lives. A central argument is that sustainability takes on different forms at various levels of analysis, and must be studied at different geographical scales. It is also vital to account for the stakeholder (households, community, tourists, businesses and government) interactions that occur both within *and* between these different scales of analysis. The chapter concludes by stressing that many of our notions of tourism and its relationship to the development process require challenging and reworking. While significant progress is being made towards this goal, considerable work still remains to be done.

What is sustainable development?

Pearce (1988) points out that at its simplest, sustainability means 'making things last' – what is being made durable can be an ecosystem, an economy, a culture, an industry, an ethnic grouping and so on. It is, in fact, far easier to

define *sustainable* than the twinned concept of *development* (Blunden *et al.*
1995). Many organisations and agencies adhere to the World Commission on
Environment and Development's (WCED 1987) well-known definition of
SD as 'meeting the needs of the present without compromising the ability of
future generations to meet their own needs'. This involves a process of change
in which the exploitation of the natural resource base, directions of investment, technological evolution and institutional dynamics operate in harmony
to enhance both current and future attempts to meet human needs.

The WCED (1987) sought to operationalise SD through seven strategies,
which can be summarised as reviving growth; changing the quality of growth;
meeting the essential need for jobs, food, energy, sanitation and water; ensuring sustainable population growth; conserving and enhancing the resource
base; reorienting technology and managing risk; and merging the environment and economics in decision making. These strategies are, in turn, founded
on a set of underlying assumptions (IISD *et al.* 1992): that alleviation of
world poverty is possible; that environmental policies are effective tools; and
that trade liberalisation will eliminate economic barriers to SD.

For Redclift (1987, 1988) two key contradictions are inherent in this definition of SD. Perhaps most problematic is the dichotomy between those who
consider ecological criteria as the most important element in sustainable thinking
and those who view 'human progress' as of paramount importance – with the
latter grouping concentrating on the continuity of development and the maxim-
isation of economic benefits on a sustainable basis (Pearce *et al.* 1987). Thus SD
embraces the contradictory ideas that economic growth is essential and that its
benefits can be available for all, *and* that economic growth causes environment
degradation, which is damaging to all (Jacob 1994; Barrow 1995). The fundamental inequality between North and South is also downplayed, especially the
reality that it is the destruction of natural assets in the South that creates value
to supply the requirements of the North's basic needs (Escobar 1995).

Some commentators argue that for SD to be achieved, the root causes of
non-sustainability must be addressed (Dovers and Handmer 1993). Societies
exhibit unequal distributions of power and knowledge. The tendency to con-
centrate power indefinitely has been controlled to some extent by different
mechanisms such as communal structures and graduated taxes. Harman (1993),
however, stresses that these checks and balances contain no truly effective
mechanism for the redistribution of societal resources. At the same time, the
growth of the multinational corporation has only led to a further concentration of wealth and power at the global scale.

Nevertheless, while SD is clearly a concept that has its weaknesses, we
should not overlook its very real value. Although the contradictory goals of
continued economic growth and ecological and societal stability/sustainability
may never be met, the concept of SD has provided a focal point around which
different stakeholders can air their concerns and attempt to find some sort of
consensus. Indeed, Redclift (1987) argues that the very strength of SD lies in
its relative vagueness.

Tourism and sustainable development

Past attempts to theorise tourism's role in the development process have tended to downplay the environmental dimension that is central to SD. Whether one looks towards product life cycle-based approaches (Butler 1980), the precepts of an explicitly political economy framework (Britton 1982, 1991) or the rapidly evolving constructs of what Teague (1990) calls the 'new' political economy (Iaonnides 1995; Williams and Shaw 1995), one struggles to find detailed attention being paid to the impact of tourism on the condition of the natural environment and people's broader quality of life.

The dominant frameworks of the 1980s (dependency theory and the tourism area cycle of evolution) also tended to emphasise structure over agency, stressing the overwhelming pressures that face destination communities as they open up to the global tourism industry. The eventual outcome of tourism's presence is seen to be unsustainable development, with most focus being placed on the negative economic and socio-cultural impacts associated with mass tourism. Butler (1992) also argued that ecotourism and alternative tourism simply represent the 'thin end of the wedge' and will eventually lead to large-scale, inherently unsustainable, development. Writers adopting a political economy perspective argue that metropolitan corporations and market conditions will determine the pace and form of tourism development, with local actors playing only peripheral roles in the process. This leads to underdevelopment and a breakdown of the basic conditions that promote sustainability (Britton 1982; Belk and Costa 1995). In simple terms, we must be cautious about the application of these approaches to the study of tourism and its relationship to SD, because the concept of sustainability is anathema to their central thesis that the industry sets in place a vicious downward cycle of dependent and *unsustainable* destination development.

While the recent introduction of new political economy and social theory perspectives to the study of tourism has allowed us to move beyond some of the constraints associated with these earlier models (Poon 1990; Urry 1990; Iaonnides 1995), a number of weak links remain in their explanatory armour. There is, for example, an urgent need for work on economic restructuring to be linked more intimately to issues of sustainability (Hudson 1995). Such an approach in the tourism arena could allow us to grapple better with a number of important questions, including: is 'Fordist' mass tourism really more environmentally destructive (on a per tourist basis) than certain types of 'alternative' tourism? will small, flexible and networked firms thrive in the new competitive environment? and, what is the link (if any) between sustainability and the size and ownership structure of establishments?

A further problem lies in the inability of 'manufacturing-biased' theories (e.g. regulation theory) to grapple with the complex dynamics of a diverse group of service sectors, where the 'personal touch' is very often a key factor in competitive success, and where the outputs being purchased are often intangible, non-storable experiences (Milne and Pohlmann forthcoming). There are also broad concerns about the true applicability of 'paradigmatic' concepts, such as Fordism and post-Fordism, to the industry (Milne and Gill forthcoming).

Another shortcoming lies in the difficulty that these bodies of theory have in dealing effectively with different scales of analysis (Dicken 1994). Regulation theory focuses primarily on the underlying structures that allow a regime of accumulation (and its attendant modes of regulation) to form and stabilise at the national level, but it does not provide complementary analyses at the regional or community scale, or cope effectively with the relationships between these dimensions (Peck and Tickell 1992).

Given the inadequacies of these approaches, it is not surprising that much of the literature dealing with tourism and SD has adopted an empirical (and largely atheoretical) stance. A great deal of attention has focused on attempts to define what eco/alternative/sustainable tourism actually means. Unfortunately, while the concept of sustainable tourism sounds attractive, suggesting a new approach and philosophy towards an old problem, the phrase remains extremely nebulous (Butler 1992).

Much of the discussion on sustainable tourism has attempted to differentiate more sustainable activities (ecotourism, alternative or appropriate tourism) from 'unsustainable' mass variants of the industry (see Brandon 1996; Ceballos-Lascurain 1996). In many respects, these attempts appear to have been rather counter-productive. De Kadt (1992) argues that policy makers should not simply distinguish between alternative tourism, which must meet high standards of social and environmental impact, and tourism in general, the negative impacts of which may be allowed to continue. In simple terms, we cannot make any particular form of tourism more sustainable unless we understand the ways in which it dovetails with all forms of the industry. It must also be stressed that 'sustainable tourism' is not really a definable 'end point'. Thus, while it is possible to map out and manage the progression toward a *more* sustainable form of tourism development, it is not possible to describe precisely a sustainable state or condition, except in a very limited sense.

This brief review has shown that 'sustainable tourism' is no more easy to define and operationalise than the broader concept of sustainable development. Nevertheless, sustainability remains a key concept around which debates between different tourism-related stakeholders can revolve.

Towards a theoretical framework

In developing a theoretical framework that will provide a more complete understanding of the links between tourism and SD, researchers need to build upon the diversity and complexity of the industry. We need to look carefully at the global–local processes that influence tourism, and to create a clearer conception of how different scales of analysis and various stakeholders articulate with each other (Taylor and Stanley 1992). Tourism must be viewed as a transaction process incorporating the exogenous forces of global markets and multinational corporations as well as the endogenous powers of local residents and entrepreneurs (Chang *et al.* 1996). There is a need to strive for a balance between structure and agency, rather than highlighting one at the expense of the other.

Scale is especially vital because different goals and constraints (economic, ecological and social) will dominate at each level (Blunden *et al.* 1995). At the household level, a key issue is the survival of the family unit, both economically and socially (Alger 1988). At the scale of the firm, the overriding goal is the survival of the business. In both cases, many of the dominant constraints are micro-economic in nature and often revolve around the wage relation. At the locality/community/destination scale the dominant constraints can be argued to be ecological and socio-political. The underlying goal is the maintenance, for many generations, of the life-supporting capacity of the landscape and related socio-economic regulatory structures. At the regional or national level, the dominant constraints are macro-economic, and goals become a mix of leisure and transport provision, foreign exchange generation and the support of tourism-dependent communities (see Blunden *et al.* 1995).

At the international scale, sustainable tourism development is affected by a range of actors and issues. Soaring population numbers and rising travel demands place tourism infrastructure and communities under tremendous pressure. The development of international trading blocs gradually frees the flow of people across borders, while increasingly opening up the service sector (and tourism in particular) to the rigours of globalised competition. Supranational actors, such as transnational corportions (TNCs) and global financial institutions, also shape the nature and progress of tourism development.

The number of decision makers decreases, and the degree of centralisation of decision making increases, as one moves up the spatial hierarchy. What happens in a specific tourism destination is the outcome of the decisions taken by hundreds of individuals and businesses acting either independently or in networks that exhibit varying degrees of formality. Decisions developed at the national level are, on the other hand, the domain of relatively few people working through more formalised structures. Influencing behaviour and actions at the individual or firm level implies communicating with and informing hundreds of individuals; by contrast, influencing behaviour at the top of the hierarchy may be achieved through more focused efforts (Blunden *et al.* 1995).

There has been little progress in the tourism literature to date in understanding the nature of the interconnections that occur between geographic levels (Chang *et al.* 1996), nor has there been much applied research into how stakeholders influence sustainability at different scales of resolution. We require a more refined understanding of stakeholders. Concepts such as community need to be 'unpacked', household and gender relations need to be inserted more effectively into the discourse, and we need to have a clearer understanding of how processes of enterprise restructuring dovetail with the goals of SD.

A central theme at each scale of analysis is the degree to which networks, of varying degrees of formality, play a role in influencing the structure, economic success and sustainability of the industry. If networking and networks are indeed a general response to pressures in the global economy (Hansen 1992), then increased knowledge of networking activity and network formation among key stakeholders (within and between different scales of analysis) is needed. Informal household and worker networks may, for example, provide additional economic support to supplement the wage relationship, helping to sustain groups that receive only marginal levels of assistance from tourism. Destination communities and regions also rely on network formation (between businesses, between the private and public sectors) for the development of competitive and sustainable tourist products.

The chapter now goes on to look at the interrelated issues of SD, scale and networking, from the perspective of the most important actors within the tourism system – destination communities, tourists, businesses and the government. It focuses in particular on how we understand the ability of these stakeholders to influence the tourism development process, and how our conceptualisation of these influences can be strengthened.

Community

The use of the term 'community' in tourism research has grown dramatically over the past two decades, in part because of the increasing development of tourism products that commodify community cultural resources. Despite this growth, few researchers have paid much attention to defining community. Where the concept is defined, researchers usually refer to a group of people living in the same locality, with some also including a notion of ecosystem or habitat in their definitions (Murphy 1985; Jamal and Getz 1995).

The sociologist Bernard (1973) distinguishes between 'the community,' which is an aggregation of people at a particular locale, and 'community,' which is characterised by social interaction, intimacy, moral commitments, cohesion and continuity through time. Stoddard (1993) refers to the 'interdependence theory' definition of community, which requires social organisation based on:

- shared values and beliefs by the individuals in the community;
- direct and many-sided relations between individuals; and
- the practice of reciprocity.

Communal groups can also be considered to provide identity, meaning and a sense of self-worth to their members, while providing a manageable scale through which to manage day-to-day affairs. New technologies and the global reach of media are also forcing researchers to reformulate, to some degree, long-held notions that space and proximity are essential to the formation of community. As Zeldin (1995: 467) notes, the Earth is being 'criss-crossed afresh by invisible threads uniting individuals who differ by all conventional criteria, but who are finding that they have aspirations in common'. Thus, 'traditional' localised communities have new tools through which to disseminate their concerns, and may, via global networks, gain new 'community members' that can represent their interests around the world. If we are to understand better the role of the local in the global we must grapple with this evolving notion of community and be fully aware of the fact that these shifts will affect both the perception and implementation of SD.

Participation

It is possible to discern two main sub-streams in the way in which 'community' has been analysed in the tourism literature. One prioritises structures external to the community and considers local residents as largely passive forces in the development process (Britton 1982; Milne 1997). The community is seen to be 'serving' the industry's needs, rather than vice versa. This, in turn, fosters the notion that communities are helpless 'victims' in the face of an onslaught over which they have relatively little control.

The other approach emphasises local agency and sees communities and their constituent members playing an active role in determining tourism's outcomes (e.g. Murphy 1985; Drake 1991; G. Taylor 1995). Communities are seen as being capable of planning and participating in tourism development, of making their voices heard when they are concerned, and of having the capability to control the outcomes of the industry to some degree.

Unfortunately, neither of these approaches deals particularly satisfactorily with issues of scale, inclusiveness and the evolving nature of community. While a few authors have addressed the heterogeneity and complexity of community structures (Brohman 1996; Harrison and Price 1996), a range of issues need to be addressed in more depth. As Hall (1994a: 154) notes: 'many models of community planning only focus on the visible dimensions of decision making. Such a pluralist notion of power is regarded as inadequate to explain the broader political dimensions of tourism development and fails to account for how certain groups may be excluded from decision-making processes.'

Proponents of community participation in the tourism development process often ignore the well-known tendency of local elites to appropriate the organs of participation for their own benefit (Brohman 1996). Factors such as gender relations (e.g. Kinnaird et al. 1994) and race (e.g. Din 1992) will affect power structures within communities, as will the ways in which these communities are embedded in broader socio-economic, political and environmental structures. For example, it can be argued that the ability of women to influence the sustainability of tourism development increases as one moves *down* the global–local nexus towards the community and household/firm levels. Women exert influence as owners/operators of small businesses (sometimes), as voters (often) and (most often) as consumers and organisers of the household. Unfortunately, relatively little work has been conducted on how tourism development meshes with the day-to-day running of households. Where household dynamics are discussed, they are often seen as passive reflectors of (or responses to) externally imposed pressures rather than active participants in the world economic system (see Alger 1988; Mullings 1996).

The factors underlying community participation must also be reviewed with some care. Such participation often occurs only with the direct blessing of decision makers further up the global–local nexus. State governments often lack the resources to plan and monitor all aspects of community well-being and may, therefore, want to shift responsibility to the local level. At the same time, supranational lending agencies and aid donors often play a role in forcing developing country governments to incorporate local perspectives into the development process (Zazueta 1995).

Experience shows that two types of participatory framework are emerging: *sanctioned participation* – a degree of local involvement in decision making, but where government sets the agenda, objectives and themes – and *independent political organisation*, which may utilise completely different means of access to decision making, including confrontation (Zazueta 1995). To date, the bulk of the documented research on local participation in tourism projects has focused on the former set of approaches.

Community participation can also be a double-edged sword (Drake 1991). While such approaches promote mutual responsibility between state and locals, incorporate vital local knowledge into projects, and provide outlets for the channelling of local political discontent, they can also be associated with a series of costs. Local participation is often expensive to run, may generate expectations that far exceed eventual outcomes, and may create new conflicts as marginal groups become more articulate and elites are able to gain a greater slice of participatory benefits through their own networks (Zazueta 1995). There is no guarantee that 'bottom-up' community-based planning in the pursuit of development will be any less piecemeal, or more sustainable on a regional or global basis, than previously adopted 'top-down' approaches. Indeed, coordination from higher levels in the spatial hierarchy may be necessary to avoid the problems associated with unilateral actions.

The tourist

If even modest environmental goals are to be achieved on the path towards sustainable development, a substantial proportion of the world's population must change their attitudes and behaviour (Barrow 1995; Redclift 1995b). Unfortunately, we know very little about the willingness of tourists to adopt some of the changes required for the achievement of more sustainable forms of tourism development.

In recent decades, market researchers have made substantial efforts to describe the characteristics of the environmentally concerned consumer. A range of descriptors, such as demographics, attitudes and personality traits, have been considered and shown to exert varying degrees of influence over final consumption patterns (e.g. Roberts 1996). While some attempts have been made to analyse consumer demand for more environmentally friendly and sustainable forms of tourism, these have generally been based on small sample sizes and specific case settings (Vincent 1995). Most researchers have, instead, decided to focus their attention on describing the characteristics of the rather poorly defined 'eco' or alternative segments of the marketplace (e.g. Eagles 1992; Milne *et al.* 1997). Much of this research tends to reveal that the average ecotourist is better educated, relatively well-off financially and somewhat older than the average 'mass' tourist. Unfortunately, we still do not know much about whether these types of tourist actually travel with the intention of minimising the negative effects of their presence and maximising the positive. We also know relatively little about the factors that individuals weigh up when evaluating which ecotourism products to purchase.

Tourism researchers appear to have much to gain from engaging with the recent 'geography of consumption' literature that has been emerging in recent years (Jackson 1995). This work stresses the fact that we, in part, define ourselves by what we purchase and by the 'meanings' we ascribe to the goods and services we acquire (Jackson and Thrift 1995). Consumption processes do not just occur prior to and during the actual tourist experience, their impact is also felt long afterwards.

An understanding of how travel experiences influence the long-term behaviour of tourists will be of considerable interest to the communities and businesses that comprise the lower scales of the global–local nexus. The latter will want to know which form of tourism experience generates the most return visitation, and which follow-up marketing strategies may be most appropriate. Communities may want to relay specific types of information to tourists in the hope that they will disseminate this to a broader audience.

Finally, it must be stressed that we still know relatively little about the specific impacts associated with different types of tourist. While we have a good sense of how nationalities and age groups behave, it is much harder to find broad-based studies that link expenditure and other characteristics to pyschographic variables. It is this type of economic, environmental and sociocultural information that is needed if communities are truly to have a greater say in the development of more sustainable forms of tourism.

The corporate dimension

Enterprises of all sizes and types help to shape the perceptions and behaviour of individual tourists, influence resident quality of life through the wage relation and the use of local resources, and often attempt to influence public policy. An understanding of the ways in which firms operate is, therefore, central to any attempt to develop more sustainable forms of tourism.

There is certainly considerable evidence that the global tourism industry is becoming more concerned with its relationships with the environment (see Healy 1996). Some commentators even argue that heightened environmental and cultural awareness on the part of the industry is a key element in the emergence of a 'new' managerial 'best practice' in tourism (Poon 1993). While corporate efforts to improve environmental performance have been documented, the degree to which companies are embracing some of the broader tenets of SD, such as justice and equity, is, however, questionable. For example, the World Travel and Tourism Council (WTTC) (see WTTERC 1993; also see WTTC website) environmental guidelines (Table 4.1) make no mention of attempting to maximise local economic linkages, or of concrete ways in which direct economic benefits may be redistributed to those who are having to bear the costs of specific tourist developments (see Eber 1992).

The issue of paying lip service to achieving SD objectives becomes even clearer in a recent World Travel and Tourism Environment Research Centre

Table 4.1   World Travel and Tourism Council (WTTC) environmental guidelines

- identify and minimise product and operational environmental problems;
- pay due regard to environmental concerns in design, planning, construction and implementation;
- be sensitive to the conservation of environmentally protected or threatened areas, species and scenic aesthetics, achieving landscape enhancement where possible;
- practise energy conservation;
- reduce and recycle waste;
- practise freshwater management and control sewage disposal;
- control and diminish air emissions and pollutants;
- monitor, control and reduce noise levels;
- control, reduce and eliminate environmentally unfriendly products;
- respect and support historic or religious objects and sites;
- exercise due regard for the interests of local populations (including history, traditions, culture and future development needs); and
- consider environmental issues as a key factor in the overall development of travel and tourism destinations.

Source: WTTERC 1993; also see WTTC website.

(WTTERC) study of the environmental performance of large tourism companies (1993). While many of the companies surveyed had constructed 'mission commitments' relating directly to SD, only approximately 10 percent had actually set targets that could be monitored. Not surprisingly, firms performed far better in areas that are easily measurable and that generate direct cost savings (energy consumption, waste management and water conservation).

It would be useful to have a clearer sense of the characteristics that make some businesses more sustainable (or willing to subscribe to the SD ethos) than others. While it seems clear that small, locally owned businesses tend to generate the most localised economic linkages, and that local operators are more likely to have a broader perception of how their actions affect local quality of life, these make up only a part of the broader sustainability equation. It could well be, for example, that larger corporations are better able to implement costly, but eventually cost-saving, environmental technologies than some of their smaller counterparts. Managers with international exposure are also, perhaps, more likely to be aware of the broader environmental and socio-cultural impacts of corporate activities than local operators.

Our understanding of the tourism industry's interactions with the SD process must be predicated on a detailed knowledge of the industry's current structure and how it is likely to evolve. In particular, we must strive to bring more of the evolving literature on networks and alliances (e.g. Hansen 1992) into the study of the tourism industry. How, for example, do networks between tourism companies form within destinations? And what role do factors such as spatial proximity play in determining their success? In addition, what role can network formation play in improving the competitiveness of small, locally owned firms that may be excluded from global travel distribution systems (Milne and Gill forthcoming).

We also need a clearer picture of the interactions that occur as companies transcend various scales in the global–local nexus. Key questions here are how much control do TNC head offices exert over local operations, and how much autonomy over policy setting do local branches actually have? There is also the important issue of how much control nation states can exert over global corporations. While some commentators argue that TNCs are inherently 'placeless' and undermine the regulatory role of the nation state, others (see Dicken 1994) argue that a TNC's domestic environment remains fundamentally important to how it operates, and that these companies are produced through a complex process of embedding in which the cultural, political and economic characteristics of the national home base are vital. We also still know relatively little about the interaction between TNCs (and smaller enterprises) and the communities that are affected, in some way, by their business activities. It would be particularly enlightening to see tourism research emerging that allows community and corporate perceptions of shared resources to be directly compared.

Whether we are discussing a TNC or a small owner-operated business, it is clear that many of the broader elements of the sustainable development equation will have to be sold to the tourism industry with a 'carrot and stick' approach; the carrots are the new opportunities that SD will create globally, and the premise (as yet very much unproven) that growth does not require a trade-off with the environment. The stick is that companies that are not proactive will be regulated into conformity. It is to the regulatory context that we now turn.

The regulatory context

Many commentators argue that market forces themselves are incapable of resolving issues related to long-term sustainability and the distribution of costs and benefits generated by tourism (e.g. Nelson 1993), and that this state of affairs is only exacerbated by the highly polarised nature of communities and the inequities extant in the global system (Brohman 1996). Others, such as the WTTC, argue that the role of government is to regulate certain macro-scale issues, while leaving the day-to-day running of the economy as unfettered by regulation as possible.

Whichever perspective is adopted, there is clearly an urgent need to incorporate visions of a more sustainable tourism into practical political solutions. These solutions will, however, be difficult to formulate until we have a clearer understanding of how legal and institutional changes hinder or assist groups that wish to engage in political action over SD issues; how the recomposition of power relations at different scales affects the priority given to more sustainable tourism management; and, how struggles over tourism-related resources shape the paths of different social groupings (Redclift 1995a).

In addressing these issues, a true sense of the interplay that exists between policy from above and from 'below' (in the form of local government and other regulatory initiatives) is needed (Drummond and Marsden 1995). Researchers must also grapple with the weight to be attached to human agency and broader social and natural structures as they attempt to determine the political processes through which the environment is managed (Redclift 1995a).

To date, much of the attention paid to the regulatory dimension in tourism development (sustainable or otherwise) has centred on the nation state (see Hall 1994a). This focus is justified as the nation remains the primary site for securing cohesion in socially divided societies and is still the most significant locus of struggle between competing forces at all scales on the global-local nexus. While some commentators argue that the role of the nation state has been reduced due to the growth of the TNC and the shift away from market intervention (see Alger 1988), others argue that the retreat of the Keynesian welfare state should not be seen as a sign of wholesale retreat by government but rather as leading to a qualitative reorganisation of its structural capacities and strategic emphases (Mullings 1996).

It is, of course, also necessary to engage with other scales of governance in the global-local nexus. Global economic institutions such as the World Bank play a significant role in funding and shaping tourism development in many of the world's poorer nations, while externally imposed structural adjustment policies create an environment that emphasises foreign exchange generation and the freeing up of inward foreign investment flows. Trade bloc formation not only eases the flow of goods, services and visitors between nations but can also lead to the establishment of pan-national environmental bodies that can have a direct say in the way in which the tourism industry utilises the natural resource base (e.g. NAFTA's Commission for Environmental Cooperation). The World Tourism Organisation, the United Nations Development Program and a number of other bodies also continue to play an important role in the formulation of tourism development strategies in emerging destinations, and in the design and provision of tourism industry training programmes.

Governance at the local and regional scales must also be addressed. There is clearly a need to understand local modes of regulation and growth better, and to grapple with the tangled hierarchy of regulatory practices that influence the local scale. In particular, it is vital to move beyond the tendency to present local government in a somewhat homogeneous fashion, and to account more effectively for the fact that these forms of government vary both between *and* within nation states. There is, for example, marked variation in the extent to which local governments have tax-raising powers and in the depth of the tax base they can draw on (Jessop *et al.* 1996).

If tourism is to contribute to the broader social, political and economic goals of SD, institutional mechanisms need to be put in place that will facilitate the participation of local residents in tourism planning (Brohman 1996). It must be remembered, however, that local strategies and decisions are seldom, if ever, made in an autonomous fashion. While there is clearly scope for political agency in the local domain, this will be systematically structured by wider economic and regulatory relations.

Any discussion of the regulatory context cannot ignore the increasingly vociferous calls from the private sector for a self-regulatory approach to be adopted towards tourism development. Proponents such as the WTTC (WTTERC 1993, WTTC website) see self-regulation as leading to environmental improvement without the imposition of the red tape and expense of regulatory control, with the tourism industry taking responsibility for its own actions and enforcing order through membership of trade associations. While the WTTERC (1993) sees a role for international/national regulation of single issues such as emissions or noise, and agrees that planning procedures are necessary when dealing with land use and building regulations, it cautions that such procedures will not be effective when planners seek to control the nature of products, customer segments or the structure of industries. The problem remains, however, that although most tourism enterprises are privately owned, many of the resources that they rely upon are 'common pool resources' such as coastland, water resources, parks and highways (Healy 1996). Thus the extent to which a call for self-regulation can be turned into workable practice is clearly open to question.

Government, perhaps more than any of the other stakeholders reviewed in this paper, has the potential power to shape the face of how tourism is promoted, planned, managed and regulated (Brandon 1996). The problem is to find the correct mixture of market orientation and state intervention that can lead to more sustainable forms of development, and then to devise a set of institutional and organisational arrangements that are compatible with this particular mixture (Brohman 1996). As Redclift (1995a) stresses, we need to consider the institutions that we bequeath to future generations, not just the state of the environment that we hand on.

Conclusions

This chapter has raised a number of important issues relating to the links between tourism and sustainable development. While 'sustainable development' and 'sustainable tourism' are difficult concepts to define, they should by no means be viewed as redundant. In fact, they provide some essential common ground for the development of dialogue between different stakeholders, who often hold divergent perspectives on the development process. At the same time, sustainable tourism should not be viewed as an end-state but rather as an 'ideal' towards which we can aim. One critical task that has already been embarked upon is the development of indicators that can help us to measure progress towards, or movement away from, the goal of more sustainable tourism development (see Ecotourism Society 1993; Nelson 1993; Murphy 1994). In attempting to achieve more appropriate forms of tourism, it is also essential that we steer away from creating a dichotomy between 'alternative' and 'mass' forms of tourism. Such a division serves little real purpose and diverts our attention away from the interlinked nature of all types of tourism development.

In attempting to come to terms with tourism's interactions with the SD process, researchers require a theoretical framework that embodies the complexity of the industry, a framework that can deal with the macro and the micro, structure and agency, the multinational corporation and the community, the biosphere and the local resource base, supranational organisations and the individual household. Our aim must be to gain a better understanding of how the different stakeholders in tourism's global–local nexus view the sustainable development process, and how their perceptions are influenced by the broader socio-economic and environmental structures within which they are embedded. It is hoped that the tentative framework outlined here will act as something of a starting point from which to begin the arduous task of constructing such a theory or, at the very least, will stimulate further debate about the complex relationships that exist between tourism and sustainable development.

Acknowledgments

The author would like to thank John Hull, Frank McShane, Pascale Thivierge, Steven Tufts and Suzan Woodley for their research assistance and acknowledge the funding received from Canada's SSHRC, Quebec's FCAR and the Auckland Institute of Technology Foundation Fund. The usual disclaimers apply.

Chapter 5

Tourism and the environment: sustainability and economic restructuring

Alan Williams and Gareth Shaw

The absence of social theory from the debates concerning sustainable development and societal management of the environment has been noted by a number of commentators (Turner 1993a; Benton and Redclift 1994; Taylor 1995; Dickens 1996). In particular, there has been a failure to link the debates on sustainability with those on economic restructuring, especially the application of regulation theory to understanding the causes and consequences of the shift from Fordism to, what remain contested interpretations of, post-Fordism. Gibbs (1996: 1, emphasis added) writes that 'Within this agenda, environmental factors have played a minor role and research into economic restructuring has effectively proceeded *in parallel* with environmental research.'

There are parallels with research in tourism, the main difference being the underdevelopment of theoretical work on tourism restructuring. This paper draws on the findings of two recent European research projects in order to investigate some of the linkages between these tourism literatures. The first of these is a study of *European Tourism: Regions, Spaces and Restructuring* (Montanari and Williams 1995) supported by the European Science Foundation as part of its Regional and Urban Restructuring in Europe project (see the special theme issue of *European Urban and Regional Studies*, vol. 3, no. 2, 1996). The second is the production of a workbook, *Tourism, Leisure, Nature Protection and Agri-tourism: Principles, Partnerships and Practice* (Williams and Shaw 1996), financed by European Partners for the Environment.

As a starting point, there is a need to consider why there has been a failure to bridge the gap between what may be seen as parallel literatures concerned with the relationships between tourism and restructuring, and those that consider the principles and practice of sustainable tourism strategies. In large part, this lack of communication stems from the poorly developed conceptualisation of both subject areas, but more particularly of the debate on tourism and restructuring. However, what we may term 'a dialogue of the deaf' also arises from the ideological and methodological differences of the two areas of research. The literature on tourism and economic restructuring emphasises changes in the organisation of production, the cultural nature of consumption and global–local relations (Britton 1991; Williams and Montanari 1995; Urry 1996). The focus is the analysis of mostly medium-term processes, with an emphasis being placed on the contingencies of place. In contrast, the literature on sustainable tourism is characterised by advocacy, is forward-looking and emphasises the opportunities for reshaping tourism production and consumption by adopting one of several contested theories of sustainability (Coccossis and Nijkamp 1995; Croal 1995; Bramwell *et al.* 1996). While there have been several significant attempts to develop the theorising of sustainable tourism, much of the literature tends to focus on the design and implementation of policy initiatives, or on case studies of individual firms and local communities.

This chapter argues that the debate on sustainable tourism could be considerably enriched by developing the links to the wider debates on economic restructuring. The micro-scale concerns of, and/or advocacy role of, much of the debate on sustainable developments can usefully be placed in the context of some of the economic and social processes that are shaping the reorganisation of the production and consumption of tourism. We also recognise that such a perspective does not deny the need also for the literature on restructuring to take into account the theoretical debates on, and the incipient implementation of, sustainable tourism strategies. However, within the confines of this chapter, we focus primarily on the need to inform the sustainable tourism literature with an awareness of the debates on restructuring, and illustrate this with reference to selected themes, drawing on the published and unpublished findings of the groups of researchers who were involved in the two European projects previously referred to (Montanari and Williams 1995; Shaw and Williams 1996). In this context, brief consideration is given to five themes:

1 intra-generational equity;
2 shifts in the mode of production – the end of the model of mass tourism?
3 new forms of consumption;
4 the organisation of production; and
5 the limits to state interventionism.

Intra-generational equity

Social considerations have often been overlooked in the debates about environmental management, and Hudson (1995: 49) argues that there are 'grave dangers in examining possible changes to more ecologically sustainable forms of production without full consideration of either the social conditions that this presupposes or the implications of this for economic and social sustainability'. Similarly, Wiessman (1994: 2) argues that sustainability 'has no meaning in conceptual terms unless it is associated with a definable reference quantity . . . defined in terms of scales of values'. He comments further that 'Sustainability becomes truly meaningful only if it is associated with specific social conditions – evaluated in sociopolitical terms – and with maintenance of values over the long term.' One of the key dimensions of these scales of values is the notion of equity.

The United Nations World Commission on Environment and Development (1987: 43) encapsulated notions of equity in its statement that sustainability 'meets the needs of the present, without compromising the ability of future generations to meet their own needs'. The theme that equity involves distributional questions both between and within generations has been taken up and reiterated in much of the subsequent debate on sustainability; for example, see Tourism Concern/World Wildlife Fund (1992). Turner (1993b) goes so far as to emphasise that because poverty is transmitted inter-generationally, concern for the legacy of future generations dictates that we must address intra-generational inequalities among the current generation. In practice, much of the socio-political debate has focused on inter-generational equity, not least because of the political difficulties of advancing the cause of intra-generational equity in the face of a rising tide of neo-liberal economic and political ideology and practice in Europe.

The neglect of intra-generational equity considerations is, however, critical given that this underlies, usually implicitly rather than explicitly, much of the discussion of such issues as partnership, community participation and the distribution of economic and environmental benefits. This raises the question of how such inequalities are socially constructed. Inequalities cannot be collapsed into one or a few social dimensions, particularly in a macro-region as diverse as Europe. Sayer and Walker (1992: 13–15) write that 'we cannot expect societies to break neatly along the fault lines of class, gender. . . . While the structural cleavages, like continental plates, run deep and grind mercilessly, they nevertheless turn and twist against one another, against circumstance, against the wilful intransigence of human beings, and against their own internal contradictions.' Nevertheless, while recognising these complexities, and the relations between agency and structure, it is possible to assert that some social cleavages have particular significance in structuring intra-generational participation in, and capacity to benefit from, sustainable tourism programmes. Income, class, race, gender, age and disability are all obvious axes around which inequalities are structured. By way of illustration, we focus here on one such dimension, gender.

Despite some progress in women's participation in wage labour and in their legal rights, there are still glaring gendered inequalities in incomes, power and influence. In addition, some of society's most powerful regulatory mechanisms are grounded in the relations of civil society and the family. Socially constructed expectations as to the roles of mothers and wives still largely determine the domestic division of labour in many, perhaps most, households. This socially constructed gender division of labour has important implications for women's participation in the waged labour market (Dex 1985). The importance of these 'dual career' relationships is particularly marked in tourism. It is not by accident that women, compared with men employed in tourism, tend to be peripheral rather than core workers, to be in part-time jobs and to receive lower wages. To some extent, this is the outcome of women's career breaks, or need to work part time, in response to the societal construction that women should care for children and families and should therefore have dual careers.

In addition, 'women workers carry into the workplace their subordinate status in society at large. The work of women is often regarded as inferior or unskilled, simply because it is undertaken by women. The definitions of skills may be no more than a social classification based on gender' (Shaw and Williams 1994: 150). This is most obvious in the way that women often perform tasks such as cooking, bed making and cleaning in small hotels and farm tourism establishments, thereby replicating the tasks they perform in, and the subordinate status that they may hold within, the domestic division of labour.

Dickens (1996: 182) emphasises that much of the debate about tourism and the environment, part of what he terms 'green utopian thought', has neglected the importance of divisions of labour, in terms of both gender and other social differences. He writes that:

> . . . one of the problems of much socialist and indeed sociological thinking has been the outright dismissal of environmental thinking rather than an active engagement with it. The result has been a ducking out of a potentially creative tension. Green utopian thought has much to offer. The issue is how to combine it with the kinds of analysis adopted by socialists . . . the issue is how to combine ecological analysis with an understanding of human society, and especially one which includes divisions of labour and the problem of knowledge as a central consideration.

Sharper focus is given to this debate by Jackson (1994: 123), who argues that 'conservation technologies are not inherently favourable to women, let alone synergistic with their gender interests'. In particular, 'households are composed of individuals with differing decision-making spheres'(*ibid.*: 122). The key points here are that sustainability strategies should, but rarely do, take into account the particular and different circumstances of women and men. This applies to almost all aspects of such strategies. Access to tourism and leisure time and spaces is gendered, as is participation in sustainable tourism strategies. The types of activity developed – with their implications in terms of seasonality, and levels and types of jobs generated – are likely to have very different implications for men and for women. Therefore, without an appreciation of how processes of restructuring are reshaping the wider roles and divisions of labour affecting men and women (and the ways in which these are intertwined with class and age), it will not be possible to have a full understanding of how intra-generational inequalities are constructed and may be modified by sustainability agendas.

Shifts in the mode of production – the end of the model of mass tourism?

According to one line of thought, there has been a major shift in the nature of tourism consumption and production, with mass tourism stagnating and perhaps even declining, and there being more rapid expansion of more individualised and flexible forms of tourism. This argument has been most cogently theorised by Urry (1990, 1996). Mass tourism, as a form of mass consumption, is characterised by standardisation of production and products, with control resting in the hands of producers rather than consumers, and frequently with very small numbers of producers. In contrast, the more individualised and flexible forms of 'new' tourism are based on place differences, and more power supposedly rests in the hands of (some groups of) consumers. In addition, tourism flows to particular places are smaller in scale, and many of the problems associated with 'massification' are avoided.

This raises the question of whether the shift away from mass tourism creates conditions for facilitating the introduction of sustainable tourism. The initial answer seems to be positive. Gibbs (1996: 5), echoing Lipietz (1992), summarises the relationship between mass production and the environment:

> Industrial mass production not only required mass purchasing power and hence a Fordist system of labour and remuneration, it also demanded massive supplies of raw material and energy from the global economy. The reification of social relations – where people relate to one another with money and commodities on the market – causes natural constraints on production and consumption to disappear from the consciousness of society.

It is not difficult to find parallels with the way that the environment has been perceived, valued and consumed in mass tourism, whether this be in the Alps or the Mediterranean regions (e.g. Gratton and van der Straaten 1994). However, it is not possible to extend this argument in any simplistic way to conclude that the demise of mass tourism and its replacement with new forms of tourism will lead inevitably to sustainability, or even to less environmentally damaging forms of tourism activity. There are two major points to be stressed here: first, that the nature of massification needs to be debated rather than asserted, and second, that the argument concerning changing modes of production and consumption is necessarily oversimplified.

First, discussions of mass tourism (see Shaw and Williams 1994) frequently do not draw a distinction between the presence of massive numbers of tourists in particular regions and their concentration in mass resorts. There are in fact two opposing views here. On the one hand, it can be argued that concentration overwhelms local environmental systems and the capacities of local societies to manage these; as a result, mass tourist resorts can be seen to be the antithesis of sustainable tourism. On the other hand, it can be contested that greater spatial diffusion of tourism would generate more travel and would bring more local systems under threat, thereby diluting the economies of scale available in mass tourism resorts in marshalling resources to deal with the environmental challenges of mass tourism. In other words, mass tourism has very different environmental implications according to how it is implanted within regions, and the contingencies of space and place.

Second, the argument about changing modes of production and consumption of tourism has tended to be oversimplified in a number of ways. The first point to emphasise is that mass tourism coexisted with more individualised forms of tourism even in the 1960s and 1970s; it was often only marginally the dominant form of tourism, and perhaps in many or even most European regions could not even claim to hold this position. The supposed shift away from mass tourism has also been exaggerated, not least because of the focus on short-term fluctuations in demand, particularly in Spain in the early 1990s. Despite the much-vaunted decline of mass tourism, there is at best only evidence of relative decline, when contrasted with some other forms of tourism. Moreover, the prediction of the early death of mass tourism is based on analyses of northern European markets and ignores the potential for the growth of mass tourism within the domestic markets of southern Europe, and more especially within Central and Eastern Europe (Hall 1995).

It may also be the case that the implications of the growing demand for more individualised, flexible and higher-quality tourism may have been mis-interpreted. Not least, there are signs that mass tourism and mass tourism resorts may be able to adapt to such demands, offering greater flexibility. This may be facilitated by increased self-provisioning, and more individualisation as a result of offering the consumer more choice in designing his/her holiday package; the latter may be supported by the increased range of choices that can be offered to customers by the use of increasingly sophisticated information technology (IT) systems in travel agencies, tour operators and hotels.

In summary, this review indicates the need to question many of the assumptions surrounding changes in mass tourism and the emergence of new forms of tourism. The debates in the restructuring literature concerning the nature of the shifts from Fordism to an unspecified form of post-Fordism (Amin 1994) could provide an essential starting point for any such analysis.

New forms of consumption and new opportunities for sustainable tourism?

Tourism is a dynamic product and, unquestionably, there are new forms of consumption, sometimes labelled post-modernist (Urry 1990). In terms of the debate on sustainability, two interlinked processes are important here (Lash and Urry 1994: 297–8). The first is the growth of consumerism, which has contributed to the critique of environmental management. In particular, this is because 'there is heightened reflexivity about the places and environments' that are consumed as a result of social encounters or visual consumption, both of which are central to tourism. 'As people reflect upon such consumption they develop not only a duty to consume but also certain rights, rights of the citizen as consumer. Such rights include the belief that people are entitled to certain qualities of the environment, of air, water and scenery, and that these extend into the future and to other populations' (*ibid.*: 297). In addition, there have been changes in the social construction of nature, such that nature and nature construction have been allocated greater value, at least among some social strata. There is little difficulty in accepting that such social processes have led to increased interest in new forms of nature-based tourism, and that this has fuelled the rise of the sustainable tourism movement. The danger lies in the temptation to conclude that new forms of consumption, in general, will lead to greater sustainability precisely because they are not mass tourism. In reality, some of the characteristics of 'post-modernist' tourism, such as de-differentiation, do not in themselves necessarily have clear implications for sustainable tourism agendas.

In practice, there are a variety of forms of new (or reinvented) tourism products, and these have vastly different implications for sustainable development. Some forms of 'new' tourism represent new forms of mass tourism. For example, mega-events (Carreras i Verdaguer 1995) can be seen as a very extreme form of mass tourism; they offer a largely standardised product whose marketing and production is controlled by a relatively small numbers of companies working in partnership with public authorities. Mega-events involve particularly intense spatial and temporal polarisation. On a *per capita/per diem* basis, the environmental challenges of, say, the Atlanta Olympic Games are probably much more acute than those of, say, Torremolinos. This is, admittedly, an extreme example, but mega-events have become increasingly prominent in Europe. A less extreme but more prevalent challenge is provided by other forms of 'new' tourism, such as urban cultural and heritage tourism, which generate intense environmental pressures in cities such as Venice (Muscara 1995). Rural tourism can also become mass tourism, as is evidenced by the overloading of environmental systems in Tuscany, the Dordogne and other bucolic landscapes that attract large numbers of tourists each summer.

One of the hallmarks of the new forms of tourism is the favouring of short breaks, brought about by increases in the availability and flexibility of free time and in disposable incomes. The role of tourism as a positional good has also favoured the growth of second and third holidays. For example, a long-haul summer holiday, a week's skiing in the Alps and two or three short breaks in Paris, Bruges and the Cotswolds or the Dales are probably not untypical of the vacation 'portfolio' of the British upper middle classes. Such tourism generates particular environmental challenges, as it will involve increased travel (much of it by air) relative to the number of vacation days. While such short breaks may help to spread the season and, therefore, are economically attractive to local operators, they present additional environmental challenges if they are complements to rather than substitutes for the existing seasonal peaking of tourism in these areas.

Tourism is a positional good and, therefore, it is subject to the dictates of fashion, which, in general, have led to shorter product cycles. As a result, investments in tourism infrastructure at particular sites may increasingly be rapidly abandoned (with corresponding new investments being required in other locations), or there is a requirement to renew the investments in these sites at increasingly short intervals. Not only is this resource-intensive and likely to lead to the 'concretisation' of more sites, but it also means that there are pressures to maximise investment returns over relatively short time periods. Such short-termism does, of course, run counter to emphasis in sustainable tourism on a long-term perspective.

The organisation of production matters

The literature on sustainable tourism has paid scant attention to the organisation of the production and delivery of tourism services. Tourism is a remarkably varied economic sector with important sectoral and locational differences in its organisation. This can be illustrated by the ways in which tourist attractions and hospitality services can be either public or private goods. In addition, tourists require a variety of transport, accommodation, catering, and other services and goods, which can be either self-provided or provided by establishments with a diverse set of ownership and operational characteristics. Tourists may also organise their own holidays or may rely on intermediaries such as tour companies and travel agents. A number of important implications for the debate on sustainable tourism follow from these differences.

Perhaps the most important point to emphasise is that the industry is highly fragmented. While some commentators hold that small (particularly dispersed) firms tend, of necessity, to be more likely to be concerned with quality and environmental issues, this is a view that we would contest. Not least, smallness is often associated with fragmentation, which is problematic given that sustainable tourism presupposes a holistic approach. Individual enterprises may introduce comprehensive green audits and programmes, but their power to affect the surrounding area and the way that tourists behave is highly circumscribed. This is problematic because sustainability measures are most likely to be resourced where they bring direct benefits to individual capital (e.g. reduced laundry bills and energy conservation). In contrast, key pressure points, such as the environmental damage caused by the car traffic generated by tourists, is very difficult to tackle not only because it is a public good but because the fragmented structure of the industry requires high levels of coordination between groups with *apparently* contradictory aims.

Ownership of the tourism industry in Europe tends to be very small scale compared with, say, the USA. For example, in the hotel industry 77 percent of upper and mid-level hotels in the USA belong to larger chains, while the corresponding figure in Europe is only 30 percent (Pizam and Knowles 1994). While sustainability is often popularly associated with 'smallness' and there are important examples of independent hotels setting high standards in terms of sustainability (Becker 1995), the link between scale and sustainability has not been empirically (or theoretically) verified. In contrast, there is evidence that sustainability projects can be diffused quickly across large numbers of branches, and across international frontiers, once they have been incorporated into the goals and business plans of major corporations such as Swissair (Wyss and Keller 1992) or Scandic Hotels (Williams and Shaw 1996).

Some sectors of the tourism industry are dominated by one particular form of organisation, the tour companies; this is particularly true of international mass tourism from northern to southern Europe but does not necessarily apply to all forms of mass tourism. The environmental implications that follow from this are mixed. On the one hand, the oligopsonistic powers of the tour companies in relations to their subcontractors, combined with sharp market segmentation, has allowed them to force down costs (Williams 1995). As a result, subcontracting firms find that they are under constant pressure to reduce their costs and, therefore, they tend to lack capital for reinvestment, which limits their potential for investing in some types of sustainable project. Given the high degree of product standardisation, there is also the implication that competition in this market is almost entirely on the basis of prices (Williams 1995), thereby militating against the introduction of all but (short-term) indi- vidual, company-level, cost-reducing sustainability measures. However, this argument is to some extent overstated and there are, for example, important differences between, say, British and German tour companies. The greater environmental awareness of the latter, with TUI for example using its power over subcontractors in order to improve environmental management in resorts, emphasises the importance of the cultural dimension of consumption. How- ever, it is possible that in the future the role of such companies may become more constrained. At present, the markets for tour operators tend to be highly segmented along national lines, but there is no reason not to expect increased internationalisation of markets. This may intensify competition, particularly in terms of prices, with negative implications for environmental management. Nevertheless, this may be too pessimistic because the greater environmental awareness of consumers in some countries (such as Germany) may provide a buttress against the environmentalism that may be associated with such global- isation tendencies.

Organisational features are also important in that successful sustainability programmes require considerable investment in staff training. The particular nature of tourism services (they have to be produced at particular places at particular times) means that tourism labour markets are characterised by strat- egies to increase flexibility (Shaw and Williams 1994), often involving an increase in the ratio of peripheral to core employees. Firms tend to engage in a variety of short-term and temporary contracts, and given the costs and commitments of many forms of training, this militates against a human resource approach to managing staff in the interests of sustainability.

Regulation: the limits of state intervention

The French Regulation school provides one of the more promising approaches to theorising the role of the state (Dunford 1990). This is an approach that emphasises the correspondence between regimes of accumulation and modes of regulation necessary for the stability of a social system. The mode of regula- tion is the set of regulatory mechanisms that are consistent with and support a regime of accumulation. In practice, this involves 'complex relations – political practices, social norms and cultural forms – which allow the highly dynamic and unstable capitalist system to function, at least for a period, in a relatively coherent and stable fashion. . . . The State clearly has a crucial role to play in all this' (Hudson and Williams 1994: 19).

The above theoretical consideration is critical for the debate surrounding sustainable tourism. A successful approach to sustainable tourism requires a holistic approach in which the actions and interests of all major stakeholders are combined, including appropriate levels of the state (Williams and Shaw 1996). In practice, there are important constraints on state intervention – whether national or transnational – in support of sustainable tourism in Europe, and this is critical given the high level of social as opposed to individual costs and benefits in the realm of sustainability.

The first point to note is that the institutional setting for tourism policy is particularly weak due to fragmentation in the industry and weak interest group representation (Pridham 1996), so tourism competes poorly against other sectors in the contest to exert influence over the state and the distribution of resources. Taken together with the general constraints on the role of the state in Western European models of social democracy, this means that state intervention in tourism has tended largely to be limited to indicative planning, reinforced by negative land use (and other) controls, because with a few exceptions, such as national parks, the state is rarely a significant direct owner of tourism resources or tourism capital. While the state may indicate goals, and can invest in public transport and other means to facilitate particular tourism programmes, ultimately the implementation of sustainability programmes depends on private capital, which may have diverse and conflicting goals.

During the 1990s, the capacity of the state to intervene on behalf of *any* tourism programmes has been further constrained by the rolling back of 'the frontiers of the state', due to both an ideological onslaught from the right and the dictates of international competition and macro-economic strategies (focused on interest rate reductions, and public sector borrowing reductions). This has meant that, in many parts of Europe, the state has tended to withdraw from active intervention in tourism, as is evident, for example, in the proposed sales of *paradores* in Spain, the privatisation of public transport in many countries, and the scaling back of (national) regional policy.

While the public sector has been withdrawing from economic intervention at the level of the nation state, it has become more active at other levels, for example, in Europe at the level of the European Union (EU) and the local. In particular, there has been growing EU involvement with formal tourism policy since the 1980s (Robinson 1993). However, this remains poorly developed and the claim of tourism to a separate 'title' (and hence its own policy) domain within the EU remains an issue for the 1996–7 Inter Governmental Conference. There is a counter-argument that tourism is already addressed under other EU titles, such as the Environment, and that this has had positive implications for sustainable tourism, as is exemplified by the introduction of the Blue Flag scheme and the funding of sustainability projects within the LIFE programme. While this cannot be denied, it is also true that other common policies, such as the European Regional Development Fund, may have encouraged tourism developments that do not accord with sustainability principles.

The other level of increased state interventionism has been the locality, where tourism has often become the mainstay of local development strategies established to fill the vacuum caused by the withdrawal of the central state and the geographical diffusion of structural unemployment. To date, most of these strategies have been driven by economic dictates and have paid scant attention to sustainability issues, even though, especially in urban areas, tourism has been used to fund environmental improvements (see Law 1993 for the example of urban tourism). Agenda 21 seeks to shift this position, but it is too early as yet to evaluate its effectiveness. This is important because, as Lash and Urry (1994: 293) argue: 'Contemporary problems and ideally contemporary solutions are global, but also . . . they are partly local. Certain aspects of the environment are only comprehensible at the local level; indeed for many people that is all of the environment that they can challenge. It is only local action that can be envisaged and sustained.'

The difficulty is that the absence of effective global regulatory mechanisms, coupled with the weakening of the nation state, places too much reliance on the roles of the EU and the local state, both of which are also constrained in their capacity for effective intervention on behalf of sustainability programmes.

Conclusions: the political economy of sustainability

The tenor of this chapter has been pessimistic. In part, this is because of its polemical approach, although we have recognised elsewhere that there is a wealth of good practice of sustainable tourism in Europe (Williams and Shaw 1996). However, we believe that the debates on the restructuring of tourism are important for the sustainability debate. Within the confines of this short exploratory contribution, we have been illustrative rather than comprehensive in our review of the links between the two areas of debate. Nevertheless, we have been able to demonstrate some of the global constraints as well as the contingencies of place and of company structures that mediate the form, speed and success of programmes and projects that seek to introduce sustainability practices.

The essence of tourism is the way in which the global interacts with the local. For example, mass tourism emphasises a global scan for destinations for global (or at least macro-regional) markets, while some forms of new tourism seek to exploit the individuality of places. These global–local relationships are not static but are subject to a variety of restructuring processes. Attempts to make tourism more sustainable will be conditioned by the very nature of these restructuring processes, but it is equally true that the grand project of sustainable tourism will also be influenced by the processes of restructuring. The debate on sustainability must converge with those in political economy, for, as Harvey (1993: 22) argues, 'ecological arguments are never socially neutral any more than sociopolitical arguments are ecologically neutral'.

Chapter 6

Land and culture: sustainable tourism and indigenous peoples

Heather Zeppel

Worldwide, indigenous peoples are becoming more involved in the tourism industry (Butler and Hinch 1996; Price 1996). Tourism enterprises controlled by indigenous people include culture-based attractions and other tourist-oriented facilities or services. These indigenous tourism ventures are largely a response to the spread of tourism into remote and marginal areas, including national parks, reserves and homelands that are traditional living areas for many indigenous groups. Indigenous cultures are frequently the main attraction for tours visiting wild and scenic natural areas such as the Amazon, Borneo and Oceania. Native lands in developed countries are also a growing focus for indigenous tourism (Lew and van Otten 1998). Indeed, the indigenous need for deriving income from land, cultural resources and new economic ventures coincides with a growing tourist demand for indigenous cultural experiences. Environmental, cultural and spiritual aspects of indigenous heritage and traditions are especially featured in ecotourism, cultural tourism and alternative tourism markets.

This chapter will examine indigenous involvement in tourism in both developing and industrialised countries. It will compare the approaches adopted by different indigenous communities in developing and implementing sustainable tourism ventures. Case studies of indigenous tourism are drawn from the Pacific region, South and Central America and East Africa. Other 'native-owned' tourism ventures in Australia, New Zealand, Canada and the USA are also reviewed. For many indigenous peoples, controlled tourism is seen as a way of achieving cultural, environmental and economic sustainability for the community (Sofield 1993; Butler and Hinch 1996; Zeppel 1997). Opening up indigenous homelands and wider involvement in tourism, however, involves a balance between tourist needs for contact with indigenous people and maintaining environmental, social and cultural integrity. Geographic and political factors, such as remoteness or indigenous control of land and culture, also influence the likelihood of indigenous peoples developing sustainable tourism ventures.

Geography of indigenous tourism

The spread of tourism into remote areas often coincides with regions that are still the traditional homelands for surviving groups of indigenous peoples.

Tourist experiences with indigenous peoples now include trekking among hill tribes in Thailand (Dearden 1993), visiting Iban longhouses in Sarawak, Borneo (Zeppel 1995), meeting Inuit people in the Arctic (Smith 1989, 1996) and Aboriginal cultural tours in northern Australia (Burchett 1992). In geographic terms, these peripheral regions encompass mountains, deserts, polar areas and tropical rainforests. These 'at-risk' environments contain both fragile ecosystems and indigenous communities vulnerable to increased accessibility and contact arising from tourism (Harrison and Price 1996). Small islands, and small island states with indigenous populations, especially in the Asia–Pacific region, are also a growing focus for ecotourism ventures (Rudkin and Hall 1996; Briguglio *et al.* 1996). Concomitant with this global expansion in tourism has been increasing concern for sustainable tourism development, particularly tourism among indigenous peoples (Nelson *et al.* 1993; Cater and Lowman 1994; Price 1996). A key focus in this chapter is indigenous control of sustainable tourism development.

Indigenous ecotourism in the Pacific region

Most small Pacific island nations typically rely on foreign aid, agriculture, fishing, logging, and tourism for income. Several ecotourism projects have been developed as alternative tourism ventures in the Solomon Islands and Western Samoa. These are intended to provide some income for local villagers and provide an incentive to conserve tropical rainforests. Western consultants and conservation agencies are heavily involved in developing these 'local' tourism ventures (Rudkin and Hall 1996). At Nggela (Florida Island), in the central Solomon Islands, unresolved grievances between Australian managers and customary landowners led to the forced entry, burning and eventual closure of the Anuha Island resort in 1992 (Sofield 1996). In the Solomon Islands, tourism developments clearly need the support of indigenous communities for sustainability.

A responsible tourism project was also proposed for Makira Island, to the east of the main island of Guadalcanal in the Solomon Islands. This ecotourism project aimed to assist in rainforest conservation and provide a cash income for Melanesian villagers still leading subsistence lifestyles. The project was developed by the Solomon Islands Development Trust, a non-government organisation, together with two international conservation agencies, Maruia Society (New Zealand) and Conservation International (USA). An initial visit to villages on Makira Island was made in 1994 by a New Zealand adventure travel company and, in 1995, by One World Travel from Australia (Volkel-Hutchison 1996). One World Travel began organising tours promoting responsible tourism and sustainable development on Makira Island. Representatives from Conservation International, however, decided that villagers would benefit more from a regular flow of visitors and decided to include mainstream travel agencies in the project. The local benefits from community participation in ecotourism thus became secondary to the need for economic sustainability of the venture.

Western Samoa has established a National Ecotourism Programme to develop village-based alternative tourism. These sustainable tourism projects include eco-villages and eco-lodges on Manono, Savaii and Upolu Islands in Western Samoa (Sooaemalelagi *et al.* 1996). Eco-villages, such as Uafato village, have a rainforest conservation area and laws set up to protect their wildlife, retain their village customs and traditions, and participate in local tourism projects. Tourists pay US$20 per night to stay at Samoan eco-villages and share in rural life (Perrottet 1996). Visitors are encouraged to help with rural development tasks and environmental restoration projects (water supply, reafforestation). Eco-villages currently receive two small groups per month, with at most 12–15 visitors (Sooaemalelagi *et al.* 1996). These small-scale, locally owned tourism projects are meant to benefit rural villagers. While 30 homepages on sustainable tourism and ecotourism have been placed on the Internet, Western Samoa still lacks alternative technologies, community facilities or an effective marketing plan for its eco-villages.

Indigenous tourism in South and Central America

Several indigenous communities in South and Central America have also developed rainforest ecotourism projects for sustainable development. In Ecuador's Napo Province, in the Amazon Basin, 24 Quicha Indian families at Capirona independently initiated a small ecotourism project in 1991 (Colvin 1994). With a loan from a regional indigenous federation, the villagers at Capirona constructed a tourist lodge and visitor centre. Practical strategies for developing and managing ecotourism at Capirona were also prepared for the community by a research team from the University of California. For Capirona, sustainable tourism is limited by small visitor numbers, food resources and competition with local tour operators and other villages. In 1995, the village of Rio Blanco attracted 150 visitors to its ecotourism project (Schaller 1996). With a growing network of Quicha villages involved in ecotourism, there is a need for varied tourist programmes and a wider range of sustainable activities (Colvin 1994).

At Punta Laguna in Mexico, a Mayan community has initiated a nature tourism project, attracting tourists travelling among Mayan archaeological sites on the Yucatan Peninsula (Long 1990). This local tourism project evolved from the determination of one man, Serapio Canul, to protect forest areas and wildlife, particularly a troop of resident spider monkeys, from exploitation. Tourists taking a forest wildlife tour with Señor Canul left either a tip or a donation. Local tour guides also brought visitors to Punta Laguna. In 1989, a Mexican non-government conservation agency, PRONATURA, provided funds for a visitor reception area and a tourist brochure. The Punta Laguna nature tourism project has provided valuable income for a poor community lacking modern facilities. Issues for sustainability were the impacts of growing tourist numbers on forest trails and wildlife.

In Panama, the San Blas Kuna Indians, numbering 30,000 people, largely control tourism in their homeland area. Living on offshore islands, the San

Blas Kuna are known mainly for the colourful outfits worn by women, espe-cially the *mola* blouse. In the mid-1970s, Kuna opposition to a large hotel development on one island led to the forced expulsion of other non-Kuna resort and tour boat operators from the area (Howe 1982; Chapin 1990). All small hotels in the area are now run by the Kuna, with tourists flown in from Panama City in light planes. However, a plan to establish nature tourism pro-jects in a Kuna wildlife reserve of 60,000 hectares was less successful (*ibid.*). Despite the attraction of a primary rainforest area, there was no support from the Panamanian government for developing ecotourism, resulting in poor road access, a lack of transport to the park centre inland at Nusagandi and limited tourist facilities. For the San Blas Kuna Indians, cultural tourism has been more sustainable than ecotourism.

Wildlife tourism in Maasailand, East Africa

The Maasai are cattle herders, well known for their warriors with red cloaks, braided hair and jumping dances. Traditional Maasai territory in Kenya and Tanzania includes famous national parks such as Amboseli, Maasai Mara, Serengeti and Ngorongoro, which are heavily visited by tourists on wildlife safaris. Employment for the Maasai living around these parks was limited to posing for photographs and selling craft souvenirs (Bachmann 1988). More recently, the Maasai have benefited from more active participation in Kenya's lucrative wildlife tourism industry, largely based in Maasailand. Some Maasai now own lodges and game-viewing vehicles, operate 'cultural' villages, have become partners in safari companies, lease their land for camps and hotels, own service shops on tourist routes and have established their own Maasai wildlife sanctuaries and tour ventures (Berger 1996). The impacts of uncon-trolled tourism, however, include environmental degradation, population growth, higher prices and the exploitation of Maasai women and children. Other Maasai group ranches funded by the Kenya Wildlife Service combine livestock rearing and ecotourism, with open grasslands attracting wild animals from adjacent reserves (*ibid.*).

In Maasailand, the sustainable management of wildlife tourism will increas-ingly involve more equitable arrangements with Maasai communities over compatible land uses, sharing wildlife revenues, ownership of lodges, and joint ventures with safari operators. The Loita Maasai, numbering around 17,000 people, are opposing plans to turn a forest area on Loita Hills known as 'The Forest of the Lost Child' into a game reserve. In particular, the Loita Maasai wish to avoid the environmental degradation and impacts of mass tourism caused by the proliferation of lodges and safari vehicles in the nearby Maasai Mara Game Reserve. Profits from safari tourism go mainly to foreign-owned travel enterprises, while local Maasai communities bordering the Mara Re-serve continue to lead poverty-stricken lives (Carrere 1995). For the self-sufficient Loita Maasai, however, the Forest of the Lost Child provides a watershed, a cattle grazing area in the dry season and a source of medicinal herbs, and is of ceremonial significance. Instead of gazetting the area as a game reserve, or allowing safari lodges and minibuses access, the Loita Maasai wish to develop low-key tourist facilities, and activities such as tented camps, forest walks with Loita elders and visiting villages bordering the forest area to participate in Maasai daily life (Stewart 1996).

Developing indigenous tourism

The global spread of tourism into remote wilderness areas and peripheral regions is reflected in growing indigenous involvement with tourism (Price 1996). In developing countries, indigenous peoples are often presented as an exotic tourist attraction with little control over the scale and impacts of tourism development (Johnston 1990). Recent indigenous tourism projects in the Solomon Islands, Ecuador, Panama, Mexico and Kenya, however, are supported by non-government organisations (NGOs), mainly conservation, environment or aid agencies, and academic consultants (Table 6.1). The importance of obtaining indigenous support for tourism development is illustrated by the failure of the Anuha Island resort in the Solomon Islands and the takeover of local tourism enterprises by San Blas Kuna Indians in Panama. Sustainable indigenous tourism ventures in developing countries also rely on government support for community-based tourism, as in Western Samoa, and developing viable links with the commercial tourism industry. For other indigenous peoples in developed countries though, growing recognition of land rights and resource

Table 6.1 Examples of involvement of NGOs and other agencies in developing indigenous tourism

Location	Agency
Makira Island, Solomon Islands	Solomon Islands Development Trust Maruia Society (NZ) Conservation International (USA) One World Travel (Community Aid Abroad) Australia
Capirona, Napo Province, Amazon Basin, Ecuador	University of California Research Expedition Program
Rio Blanco, Napo Province, Amazon Basin, Ecuador	David Schaller, University of Minnesota, USA
Punta Laguna, Yucatan Peninsula, Mexico	PRONATURA – Peninsula de Yucatan Veronica Long, University of Waterloo, Canada
Kuna Wildlife Reserve, Panama	Smithsonian Tropical Research Institute World Wildlife Fund Agency for International Development Inter-American Foundation
Loita Maasai, 'Forest of the Lost Child', Kenya	Third World Network, Penang, Malaysia

ownership, together with the current trend for cultural revival, provides the main impetus for developing tourism. In this situation, 'appropriately managed tourism is seen as a sustainable activity that is generally consistent with indigenous values about the sanctity of the land and people's relationship to it' (Hinch and Butler 1996: 5). This nexus between land and culture defines sustainable tourism for indigenous peoples. In the next section, case studies are presented of Aboriginal tourism in Australia, Maori tourism in New Zealand, First Nations tourism in Canada and Pueblo Indian tourism in the southwest USA.

Aboriginal tourism in Australia

The Northern Territory, where Aborigines comprise 25 percent of the population of 176,000, has played a key role in the development of Aboriginal tourism in Australia. By 1996, 52 Aboriginal tours were listed in the Northern Territory's Aboriginal tourism marketing brochure, *Come Share our Culture*, produced since 1990 (Burchett 1993; Zeppel 1996). An Aboriginal tourism manager was first appointed by the Northern Territory Tourist Commission in 1984 (Burchett 1992). A decade later, Aboriginal tourism brochures are now also produced in Queensland (since 1994) and South Australia (in 1996). However, the Territory's northern manager for Aboriginal tourism believes 80 percent of all viable (i.e. market-ready, operational) Aboriginal tourism products in Australia are still located in the Northern Territory (Joc Schmiechen, personal communication).

The Northern Territory includes the well-known tourist destinations of Kakadu and Uluru (Ayers Rock) National Parks, both jointly managed with traditional Aboriginal owners (Altman 1989; Mercer 1994). With successful land claims lodged under the Aboriginal Land Rights (N.T.) Act (1976), 38 percent of Northern Territory land is now owned by Aboriginal people, with a further 10 percent under claim (Northern Territory Tourist Commission 1994). These homeland areas have become the main focus for Aboriginal tourism ventures, with negotiated access to specific cultural sites, often developed in conjunction with non-Aboriginal partners and the tourism industry. The economic importance of this new sector is reflected in the recent Aboriginal Tourism Strategy developed for the Northern Territory (Northern Territory Tourist Commission 1996).

Manyallaluk: the Dreaming place
Aboriginal groups in Australia are actively developing tourism ventures in homeland areas and other regions of continuing cultural significance. One such Aboriginal-owned and -operated tourism venture is Manyallaluk, 100 kilometres east of Katherine in the Northern Territory. This community-based Aboriginal tourism venture has won both Australian and Northern Territory tourism awards for 1993, 1994 and 1995. Based on the former Eva Valley cattle station, Manyallaluk (Frog Dreaming) is a community of 150 Aboriginal people. The rugged homeland area of Manyallaluk covers an area of 3,000 km^2,

Table 6.2 Aboriginal rules for cultural tours in Northern Territory homeland areas

---

Manyallaluk: The Dreaming Place
- Visitors are not allowed beyond the homestead area unless on a guided tour.
- Visitors are asked to respect our privacy and not enter the community area.
- Manyallaluk is a restricted area and NO ALCOHOL IS ALLOWED.

Umorrduk Safaris Arnhem Land
- Stringent conditions apply for entry to traditional land. Visitors will be issued with a copy of these conditions and will be required to indicate their acceptance by personal signature.
- Umorrduk Safari Guides will indicate where taking of photographs is prohibited.
- The traditional owners reserve the right to close particular areas for ceremonial purposes and undertake to make suitable alternative arrangements.

Source: *Manyallaluk: The Dreaming Place* brochure (valid to 31 March 1998) and Umorrduk Safaris Arnhem Land in Odyssey Safaris *Australian Outback Expeditions* brochure (valid to 30 April 1997).

bordered by Arnhem Land, Nitmiluk (Katherine Gorge) National Park and Kakadu National Park. Aboriginal tourism at Manyallaluk started in 1990, as a joint venture with Terra Safari Tours. It began by training six Aboriginal tour guides, including stockmen from the former Eva Valley cattle property.

At Manyallaluk, Aboriginal guides conduct cultural tours of one to five days where tourists gain hands-on experience in making or using traditional artefacts (basket weaving, spear throwing, bark painting, fire lighting), learn about bush food and plant medicines and visit ceremonial rock art sites (McHugh 1996). While coach tours and self-drive visitors are welcome at this community, 'the people of Manyallaluk make it clear that it is their country and they alone have the right to show it to others' (Gill 1994: 23). The enterprise employs 20 Aboriginal people, mainly as tour guides, with the community also operating a caravan park, camping ground and store selling local Aboriginal artefacts (Gill 1994). The success of Manyallaluk has derived from Aboriginal input and control over tourism development. Tourists are restricted to the homestead area unless on a guided tour and further requested not to enter the community living area (Table 6.2). The main aim of tourism at Manyallaluk is to create an independent and sustainable lifestyle for the resident Aboriginal community.

Umorrduk Safaris Arnhem Land

Umorrduk Safaris is one of only four local tour operators granted access to Arnhem Land, a large Aboriginal homeland area in the northeast region of the Northern Territory. Arnhem Land is owned by just five or six related Aboriginal clans and contains superb galleries of 'X-ray' style Aboriginal paintings (Mulchand 1993). It adjoins the World Heritage-listed Kakadu National Park, and entry is by permit only. Umorrduk Safaris operates a small tented safari camp in western Arnhem Land, on the Mudjeegarrdart homelands of the Gummulkbun people (Ellis 1994). Permission to visit this 1,000 km^2 clan area was granted in 1988 by a traditional owner and elder of the Gummulkbun people, John Williams. His daughter, Phyllis Williams, runs Umorrduk Safaris together with her husband Brian Rooke, of Tasmanian Aboriginal descent. Permit entry fees (A$25 per person) are paid to clan members as royalties (Mulchand 1993). Small groups of tourists are flown in from Darwin to the Umorrduk camp, visiting 20,000-year-old paintings in rock art galleries with an Aboriginal guide. A maximum of 16 people stay in eight twin-share tents. Umorrduk Safaris receives 400–500 tourists a year, with 70 percent coming from the USA (Ellis 1994).

All decisions on tourism development at Umorrduk are made in consultation with clan elders. This area of northwest Arnhem Land contains the most prolific and vibrant rock art galleries in the world (Gill 1994). While these stunning Aboriginal paintings are a key attraction, visitors to Umorrduk are not allowed to take photographs of ancient burial sites in rock shelters (Mulchand 1993). Umorrduk Safaris has developed as a small-scale Aboriginal tourism business to minimise environmental and cultural impacts. It attracts tourists willing to pay A$430 per person for a one-day fly-in visit with a minimum of four people. Gummulkbun clan members are employed as tour guides. Umorrduk Safaris also limits its tourism operation to eight safari tents, although the leader of the Gummulkbun clan would like to build a motel (Mulchand 1993). With a five-year lease, Brian Rooke has the final word on appropriate tourism development. Operating as a husband-and-wife company, however, Umorrduk Safaris has had no access to funding schemes for training family members as Aboriginal tour guides (Rooke 1993). In the future, the Gummulkbun clan will have the option of buying the tour business, with the present operator staying on as manager (Gill 1994). As an incorporated body, the clan will have better access to funding for Aboriginal tourism.

Sustainable Aboriginal tourism

For Aboriginal people, developing sustainable tourism requires a balance between cultural and environmental considerations, matched, to a lesser extent, by economic success (Altman 1992). With a secure land base, as in the Northern Territory homeland areas, Aboriginal groups are able to control both tourist access and the level of visitation. Aboriginal rules for cultural tours also influence tourist behaviour. Likely environmental impacts include the depletion of bush foods and wildlife, where tourists join Aboriginal hunting trips near base camps, such as on Melville Island (Burchett 1992). Smaller visitor numbers limit the impacts of tourism but require higher tour prices. Building the commercial viability of small-scale Aboriginal tour operators is a key aim of the Northern Territory's Aboriginal Tourism Strategy. Aboriginal responsibilities to maintain cultural sites also limit the size and scale of any tourism development (Burchett 1993). The conflicting demands of culture and commerce have thwarted many Aboriginal tourism ventures (Finlayson 1991). Issues regarding the appropriate presentation of Aboriginal cultural knowledge to visitors are also paramount for Aboriginal groups involved in tourism. 'In an Aboriginal way, management of tourism is firstly about managing cultural property in terms of land and resources and, secondly, negotiating the terms of trade for the use of those resources, including people' (Wells 1996: 55). Land and culture are the basis for Aboriginal tourism.

Maori tourism in New Zealand

Maori cultural heritage is a growing attraction for visitors to New Zealand (Maiden 1995; Walsh 1996). Maori carvings, action songs, *haka* and *poi* dances, the *hangi* (feast), and visits to *marae* (villages) are key attractions in Rotorua, a Maori cultural heartland. The New Zealand Maori Arts and Crafts Institute and the Whakarewarewa Thermal Reserve in Rotorua, with 70 Maori staff, attracts 400,000 visitors annually (Young 1994). More diverse Maori tourist experiences now include whale watching at Kaikoura, Maori cultural protocol at *marae*, Maori-guided tours of nature reserves, horse trekking, forest walks, and learning about the Maori way of life (Maiden 1995). The Aotearoa Maori Tourism Federation, established in 1988, actively promotes Maori tourism in New Zealand, and drafted the Maori Tourism Strategic Plan in 1995.

Greater Maori participation in New Zealand's tourism industry arose from the Maori Tourism Task Force Report of 1987. The report identified the need for government support of Maori tourism, especially tourist developments on *marae* and Maori control over tourist presentations and interpretation of Maori culture (Young 1989). Tourist interactions with Maori people in New Zealand are now mainly centred around visiting the *marae* of Maori communities, located around North Island. Maori protocol and welcome rituals have been adapted to hosting tourist groups (Maiden 1995). These Maori-guided tours also give tourists 'the opportunity to meet Maoris not only as performers in costume, but as ordinary people' (Walsh 1996: 260). Culturally sustainable Maori tourism involves Maori interpretation of diverse cultural practices, not just concert party performances (Ministry of Tourism 1992).

Maori tourism is based around access to both cultural and natural resources. Maori rights to land and resources have been increasingly asserted through the Treaty of Waitangi, originally signed in 1840, and recommendations made by the Waitangi Tribunal established under the Treaty of Waitangi Act 1975 (Hall *et al.* 1992). This includes legal recognition of Maori tribal sovereignty, land ownership and rightful usage of natural resources. Maori *iwi* (tribes) are now pursuing a range of tourism development options on tribally owned lands, including hotels and visitor centres. For example, Turangawaewae Marae plans to build a NZ$1.7 million tourist complex, including a theatre, cafeteria and souvenir shop, opposite the *marae* (*Nga Korero* 1995a).

Other Maori groups are asserting their rights over commercial access to natural attractions (Keelan 1996). In November 1996, Maori people erected a fence blocking access to the Kaituna River via Ngati Hinerangi tribal land, near Rotorua. The dispute concerned access payments from commercial rafting companies that used the Kaituna River. The river includes the highest single-drop waterfall commercially rafted in New Zealand. Recreational users were still granted access to the river landing (*Nga Korero* 1996a). On the Whanganui River, Maori owners of Kemp's pole, an 8 m pole erected in 1880 by Te Rangihiwinui, objected to a jetboat operator taking tourists to this site without asking for permission (*Nga Korero* 1996b). At Kaikoura, on South Island, the successful Whale Watch tour venture owned by the Ngai Tahu Maori mounted a High Court bid to take over control of whale-watching operations from the Department of Conservation. Operating since 1987, they use four vessels for viewing sperm whales. Another 13 operators are seeking whale-watching licences at Kaikoura. The Ngai Tahu still claim that tribal sovereignty overrides the Marine Mammals Act (*Nga Korero* 1995b). Such a claim would assure economic sustainability for Maori whale watching.

First Nations tourism in British Columbia, Canada

In British Columbia, in 1994, 182 native-owned businesses were directly involved in the tourism industry (Kramer 1994), which is double the number of native-owned tourism enterprises in 1983, with First Nations tourism developments 'the fastest-growing sector of B.C.'s tourism economy' (Zukowski 1994: 4). These new tourism ventures promote cultural experiences with First Nations people in British Columbia or involve mainstream tourist facilities such as hotels, lodges and recreation parks. Other indigenous tourism enterprises involve new partnerships between government agencies and First Nations cultural groups. For example, on Queen Charlotte Islands the Haida Gwaii Native Watchmen provide tours visiting abandoned Haida villages at Skedans and Skungwai (Ninstints) in Gwaii Haanas National Park Preserve (Halliday and Chehak 1996). Most of the other new First Nations tourism developments, however, are located on reserve lands in British Columbia.

Ahousaht 'Walk the Wild Side' Tour, Flores Island Clayoquot Sound

On Flores Island, the Ahousaht Band (1,411 members, 678 on reserve) of the Nuu-cha-nulth Nation operate a 'Walk the Wild Side' tour departing from Tofino. Located on the scenic west coast of Vancouver Island, the small town of Tofino receives half a million visitors annually (McPhedran 1995). On this native-run ecotour, Ahousaht guides explain Nuu-cha-nulth history, culture and use of natural resources while leading tourists along old forest trails and beach areas.

Flores is the largest island in Clayoquot Sound, at 16 km long and 13 km wide. The coastline of Flores has over 20 beaches, often visited by tour operators from Tofino. The Ahousaht people on Flores Island began their 'Walk the Wild Side' tour in March 1994 with an initial investment of C$158 (Jones 1995). Most of the people in the village of Ahousaht supported this tourism project, helping to clear old forest trails and building a craft store. In 1994, 600 tourists completed the 'Walk the Wild Side' tour on Flores Island and bought Ahousaht crafts. As of November 1994, the enterprise had earned C$18,000, with business expected to double in 1995 (Daniels 1995). This non-profit tourism venture, initiated by Ahousaht First Nations women, provides an alternative to the logging industry, generates local income and utilises Ahousaht cultural traditions.

'Walk the Wild Side' has benefited the Ahousaht community both econom-ically and culturally. Of the money received from the tour in 1994, over C\$7,000 was paid out to Ahousaht craftmakers, who received 90 percent of the sale price; another C\$5,000 was paid in boat fares to the Ahousaht Band; C\$2,700 to the Nuu-cha-nulth Booking Centre; C\$2,500 to various Ahousaht guides; and C\$2,500 to private charters (Jones 1995). Further money was spent locally on purchasing VHF radios, brochure printing, insurance and other ex-penses. Additional benefits for the Ahousaht include young people developing their personal abilities and confidence through dealing with tourists. Two past Ahousaht guides on 'Walk the Wild Side' now work in paying positions with Ahousaht Fisheries and at Long Beach Model Forest Society (*ibid.*).

Cultural benefits of this native ecotour include the revival of traditional crafts, with Ahousaht craftspeople teaching carving and beadwork skills to their families. Young people on Flores Island are also reconnecting with the land and Ahousaht cultural traditions. According to the tour coordinator, 'Wild Side, since its inception, has watched guides reconnect with elders in Ahousaht to learn of the oral history, language and traditional uses of the land' (*ibid.*). The preservation of Ahousaht culture, integrity and traditions has guided the local development of tourism on Flores Island. Walking trails used by 'Walk the Wild Side' tours have also been upgraded by the Ahousaht with the help of a grant received from the Western Canada Wilderness Com-mittee. Ahousaht success with 'Walk the Wild Side' has shown that the overall benefits of First Nations tourism can outweigh other traditional industries such as logging and fishing.

Cowichan Native Heritage Centre, Duncan Vancouver Island

The Cowichan Tribes in Duncan, currently numbering 2,500 people, are a Coast Salish (Hul'qumi'num) First Nations group in southeast Vancouver Island. They traditionally occupied 13 villages along the Cowichan River, with some families still living on old sites (Coull 1996). The Cowichan Band have nine reserves on 2,493 hectares adjacent to the city of Duncan and the municipality of North Cowichan. The Native Heritage Centre, located on reserve land in Duncan, was implemented by the Khowutzun Development Corporation, 'one of BC's biggest native enterprises', which is owned by the Cowichan Tribes (Daniels 1995). Launched in 1990, this C\$3.3 million her-itage tourism project initially operated for seven months, declared bankruptcy, then reopened after restructuring debts and devising a new business and mar-keting plan (Nathan 1991). In 1994, the Native Heritage Centre had sales of C\$1.2 million and employed 15 to 20 full-time staff and another 70 part-time workers, indicating a successful tourism business venture (Daniels 1995).

The Native Heritage Centre occupies six green acres on the north bank of the Cowichan River in Duncan. A tour of the centre, with a Cowichan guide, includes an introductory talk on Cowichan history and culture at a traditional fire circle; a short walk to the swift-flowing Cowichan River to view a salmon-processing site; a look inside the monumental cedar Big House relocated after Expo 86 in Vancouver, where women demonstrate wool weaving and beadwork;

a visit to the carving shed where new totem poles and canoes are fashioned; and viewing an exhibition of Cowichan-knitted wool sweaters. At the Longhouse Story Centre, visitors watch *Great Deeds*, a well-produced and very moving 25-minute film on the history of the Cowichan Nation. The Khowutzun Art Gallery, at the Native Heritage Centre, also sells a wide range of First Nations carvings, jewellery, books and art.

At the end of 1996, the Native Heritage Centre was renamed the Cowichan Native Village. Managers of the Cowichan Village have implemented a cooperative marketing scheme with four other urban-based Aboriginal tourism properties in western Canada (Quaaout Lodge, Tin Wis Resort, Luxton Museum, Wanuskewin Heritage Park) (*Canada Communique* 1996) in an effort to ensure commercial sustainability.

Pueblo Indian tourism in the Southwest USA

There are 15 Pueblo Indian tribes in the Southwest USA (Mallari and Enote 1996). In New Mexico and Arizona, the traditional multistory adobe *pueblos* (villages), along with Pueblo Indian pottery, art, jewellery, and religious traditions, have long been a major tourist attraction. Taos Pueblo, in northern New Mexico, has been visited by tourists since the early 1900s (Lujan 1993). Pueblo arts and crafts and tourism provide a major source of income for Pueblo communities. In northern New Mexico, eight Indian *pueblos* produce their own tourist brochure, covering Taos, Picuris, San Juan, Santa Clara, Nambe, San Ildefonso, Pojoaque and Tesuque (Eight Northern Indian Pueblos Council 1996). These *pueblos* feature a wide range of cultural tourist attractions and other facilities (Table 6.3). The benefits of *pueblo* tourism in providing employment and supporting community development are highlighted in this brochure. For Pueblo communities, sustainable tourism involves maintaining cultural integrity and limiting tourist access, while still gaining the economic benefits of tourism.

Taos Pueblo has been continuously inhabited for around 1,000 years. It is the northernmost of 19 *pueblos* in New Mexico and was declared a World Heritage Site in 1992 (Lujan 1993). Taos Pueblo has a tribal membership of 2,000, with 150 people still living within the four- and five-story adobe *pueblo* dwellings, without electricity or running water. Most families live in more modern homes around the old *pueblo*. Residents of Taos Pueblo deal with the constant influx of tourists by controlling and limiting access. Tourist pamphlets list fees for admission and rules of visitor behaviour. Sections of the *pueblo* are also permanently closed to tourists, while certain ceremonies at the *pueblo* are closed to all outsiders. Since 1989, Taos Pueblo has been closed to all tourists during the entire month of February for religious reasons (Lujan 1993). Despite its reliance on tourism income, the *pueblo* is closed to maintain cultural values and protect religious beliefs. At Taos Pueblo, income from tourism is used to support tribal development programmes, including education and administration.

Table 6.3 Tourist attractions and facilities in Pueblo Indian communities, Southwest USA

Community	Attractions and facilities
Taos Pueblo	Traditional *pueblo* dwelling Feast day ceremonies Shops, outdoor ovens
Picuris Pueblo	Visitor centre: shop, museum and restaurant 700-year-old *kivas* and storage rooms (Majority owner of Hotel Santa Fe)
Santa Clara Pueblo	Polished black pottery and crafts Puye cliff dwellings and Harvey House Santa Clara Canyon outdoor recreation area Special tours: Pueblo feast meal, dances, pottery demonstrations
San Ildefonso Pueblo	Black-on-black pottery Museum, pottery artisans Family-owned art and craft shops
Nambe Pueblo	Nambe Falls recreation area Nambe Pueblo tours Buffalo herd Micaceous clay pottery and crafts
Pojoaque Pueblo	Tourist information centre: Pueblo Indian artwork Poeh Cultural Centre and Museum 'Cities of Gold' Casino
Tesuque Pueblo	Camel Rock Casino, Camel Rock camping ground Tesuque trailer park, Tesuque natural farms Arts and crafts, public dances

Source: *8 Northern Indian Pueblos 1996 Visitors Guide*. Eight Northern Indian Pueblos Council, New Mexico.

The Zuni Indian community in west central New Mexico, with a population of 9,562, is centred on the town of Zuni (Mallari and Enote 1996). While Zuni artists are renowned for their crafts and jewellery, the town of Zuni is off the main tourist path. Proposals for Zuni tourism development have met with concerns over Zuni religion, protection of sacred sites, tribal sovereignty and land use. Under Zuni law, for example, tourists are not allowed to record Zuni religious events. In June 1990, the tribal government banned all non-Indians from attending Zuni religious dances. 'Tour buses are still not allowed to enter Zuni during the summer rain dances and the large winter ceremony, Sha'lako' (Mallari and Enote 1996: 27). Cultural integrity was deemed more important than tourist income, even though Sha'lako was Zuni's biggest tourist attraction, annually drawing thousands of visitors. While tourist access to back-country areas with Zuni sites is also restricted, big-game hunting by non-Indians with Zuni guides has been approved by the community. Zuni cultural concerns have determined the appropriate forms of sustainable tourism development.

Controlling indigenous tourism

Reserves and homeland areas are the main focus for indigenous tourism in Australia, Canada and the USA. In New Zealand, there is a greater focus on Maori groups claiming resource rights, including access to rivers and wildlife utilised for tourism ventures. Key factors increasing the benefits of tourism for indigenous peoples in developed countries include land ownership, community control of tourism, government support for tourist development, restricted access to indigenous homelands and reclaiming natural or cultural resources utilised for tourism (Table 6.4). Recognition of tribal sovereignty has enabled indigenous groups to control and manage tourism effectively. This includes limiting tourist access and use, establishing permitted tourist activities and the level of cultural knowledge shared (Table 6.5). Indigenous

Table 6.4 Key features of indigenous tourism in developed countries

1 Land ownership: reserves/homeland areas/resource rights
    Treaty of Waitangi Act 1975, New Zealand
    Aboriginal Land Rights (Northern Territory) Act 1976, Australia

2 Community control of tourism
    Khowutzun Development Corporation (Native Heritage Centre, BC, Canada)
    Manyallaluk Aboriginal Corporation (Manyallaluk, NT, Australia)

3 Government support for indigenous tourist development
    Northern Territory Tourist Commission, Australia
    Ministry of Maori Development, New Zealand

4 Restricted access to indigenous homelands/communities
    Umorrduk Safaris Arnhem Land, Gummulkbun Clan, NT, Australia
    Taos Pueblo/Zuni Village, New Mexico, USA

5 Indigenous tourism organisations
    Aotearoa Maori Tourism Federation, New Zealand
    Canadian National Aboriginal Tourism Association, Canada

6 Indigenous tourism strategies
    Maori Tourism Task Force Report 1987, New Zealand
    Maori Tourism Strategic Plan 1995, New Zealand
    Aboriginal Tourism Strategy 1996, NT Tourist Commission, Australia
    National Aboriginal and Torres Strait Islander Tourism Strategy 1997, Australia

Table 6.5 Indigenous control of tourism for sustainable development

Controls	Example
Spatial limitation	Hosts set limits on entry to homelands and sacred sites
Activity limitation	Hosts establish preferred or permitted tourist activities
Temporal limitation	Hosts indicate appropriate times for tourist access and use
Cultural limitation	Hosts set limits on access to cultural knowledge and rituals

tourism organisations, policies and strategies are also common features that help to represent and define native interests in tourism developments. Dimensions of sustainable tourism, however, vary for indigenous communities in remote areas and indigenous groups presenting aspects of traditional culture in urban localities. The Manyallaluk Aboriginal community in Australia and the Ahousaht Band in Canada emphasise cultural and environmental sustainability in homeland tourism. Urban-based indigenous attractions, such as the Cowichan Native Village, place a greater emphasis on economic sustainability. The key factor throughout is indigenous control of land and resources.

Conclusion: sustainable tourism and indigenous peoples

As well as being an exotic tourist attraction, indigenous peoples are increasingly also the owners and managers of culture-based attractions and other tourist facilities. Indeed, a growing tourist demand for indigenous cultural experiences coincides with the indigenous need to derive income from land, cultural resources and new economic ventures. This global trend is reflected in increasing contact with indigenous communities living in remote areas and also the opening up of indigenous homelands for tourism. Achieving sustainable tourism, though, depends on geographic location, indigenous control of land and resources and developing effective links with the wider tourism industry. In developing countries, sustainable tourism ventures for indigenous peoples are mainly implemented with the help of non-government agencies. Tropical rainforest areas in the Pacific, Central America and Africa are a main focus for these community-based indigenous tourism projects. With greater control over homeland areas, culture and resources, indigenous groups in Canada, Australia, New Zealand and the USA are directly determining the most appropriate types of tourism development. This includes controlled tourism in traditional lands and also urban-based cultural attractions. Negotiating access to tribal land, resources and cultural knowledge will be a growing feature of indigenous tourism. For indigenous peoples, environmental and cultural integrity are central for sustainable tourism development.

Chapter 7

Tools for sustainability analysis in planning and managing tourism and recreation in the destination

Pamela Wight

Tourism, at its best, is an enriching experience for the visitor. It can benefit heritage and other sites, and provide employment, income and other benefits for host communities. But, if badly planned or managed, it can turn into a disaster for the visitor, the place and the host community. However, as Gunn (1994: 83) observed, 'whereas some erosion and pollution of resources is caused by great numbers of visitors, most environmental damage is caused by *lack of plans, policies, and action to prepare for any economic growth*. . . . Tourism cannot be blamed for environmental deterioration caused by bad decisions rather than real visitor impacts.' Therefore, the relationship between tourism and the overall environment is critical: if the natural environment or the culture is damaged, or if tourism is weakened, then we lose a positive force motivating people to sustain and enrich the environment.

The concept of sustainable development is the key to seeking a more productive and harmonious relationship between the three elements: visitor, host community and environment. However, this does not imply a static relationship. Indeed, achieving harmony is dependent on the ability to accept, absorb and adapt to change, which may be large or small, expected or unexpected. This chapter examines approaches and tools that have the potential to deal with important sustainable tourism planning and management issues in the destination.

Responsible environmental practices have recently moved to the forefront of many industry agendas. In addition, a values-based perspective is being introduced. Values may be 'held' (i.e. the ideals held by an individual) or 'assigned' (i.e. the worth of something), but both are related: held values can determine preferences that function to assign relative value. For sustainability, the values that should be incorporated, whether held or assigned, relate to environment, society and economy (Figure 7.1).

There have been calls for ensuring that tourism is planned, developed and operated within the context of sustainable development principles (e.g. Australia 1991). While sustainable tourism principles are well-articulated (English Tourist Board *et al.* 1991), the 'hows' are less frequently discussed. Theorising

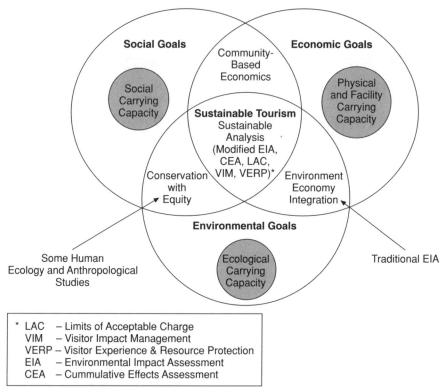

Figure 7.1   Tools to assist sustainability analysis for sustainable tourism systems (adapted from Dalal-Clayton and Wright 1994).

is useful, but tourism professionals need practical tools to implement some of the new approaches. One of the approaches regularly advocated is determining the 'carrying capacity'. Other tools include environmental impact assessment (EIA), limits of acceptable change (LAC), visitor impact management (VIM) and visitor experience and resource protection (VERP). These tools, which work in the interface between various goals and values (see Figure 7.1), are discussed below, together with some of their strengths and limitations (Table 7.1).

Carrying capacity

'Few topics in recreation management are discussed as widely or as loudly as carrying capacity. The term is a perfect example of conventional wisdom: everyone talks about managing our recreation resources within their carrying capacity; but when you get to specifics – how many, what kinds, when, for whom, etc. – the discussion bogs down' (Lime and Stankey 1971: 182). Tourism as a discipline seems to be lagging behind recreation in recognising this reality.

Table 7.1   Comparison between the elements of LAC, VIM and VERP

LAC	VIM	VERP
Initiate and identify issues		
1 Define issues and concerns	1 Preassessment data base review	1 Assemble the project team
Goals		
2 Define and describe opportunity classes	2 Review management objectives	2 Develop statements of park purposes, significance and primary interpretive themes
Standards and indicators		
3 Select indicators of resource and social conditions	3 Selection of key impact indicators	3 Map and analyse resources and visitor experiences
	4 Selection of standards for key impact indicators	4 Establish the spectrum (or range) of desired resource and social conditions (potential management zones)
		5 Use zoning to identify proposed plan and alternatives
		6 Select quality indicators and specify associated standards for each zone
Inventory and evaluation		
4 Inventory existing resource and social conditions	5 Comparison of standards and existing conditions	7 Compare desired conditions to existing conditions
5 Specify opportunity class standards	6 Identify probable causes of impacts	8 Identify probable causes of discrepancies between desired and existing conditions
6 Identify alternative opportunity class allocations		
Actions, implementation and monitoring		
7 Identify management actions	7 Identify management strategies	9 Develop/refine management strategies to address discrepancies
8 Evaluate and select alternative management actions	8 Implementation	
9 Implement and monitor		

Carrying capacity is a term borrowed from wildlife ecology, with a rather precise use – the 'maximal population size of a given species that an area can support without reducing its ability to support the same species in the future' (Daily and Ehrlich 1992: 762). But it is difficult to apply this definition to humans for many reasons, not least because in the global economy regions are not isolated. As Wackernagel *et al.* (1993) observe, economists actually regard trade flows as one way to overcome the constraints on regional carrying capacity imposed by local resource shortages.

Carrying capacity has been applied to land use planning and growth management, and other aspects of human activity. Planners have enlarged the definition of carrying capacity to include the many variables inherent in man-made systems. Shelby and Heberlein (1984) subdivided carrying capacity into ecological capacity (ecosystem parameters); physical capacity (space parameters); facility capacity (development parameters); and social capacity (experience parameters). However, because of its origins in the natural sciences, the term 'carrying capacity' suggests an objectivity and precision not warranted by its use in planning involving human systems.

A major problem in the capacity literature is that impact and evaluation often become confused. An example is the concept of 'resource damage.' All human use has an *impact*, but is it *damage*? The term 'damage' refers to a change (an objective impact) *and* a value judgment that the impact exceeds some standard. It is best to keep these two separate. In terms of human impact, a certain number of hikers may lead to a certain amount of soil compaction. This is a *change* in the environment, but whether it is *damage* depends on management objectives, expert judgments and broader public values. Most carrying capacity conflicts do not revolve around resource questions but rather around values questions (Shelby and Heberlein 1984). As Lime and Stankey (1971: 182) observed, 'determining carrying capacity ultimately requires the consideration of human values. Because of the subjectivity of these values, it is essential that managers carry on an active dialogue with a variety of publics.'

Tourism carrying capacity is frequently mentioned as one method of controlling the direction and consequences of development. Unfortunately, carrying capacity as a guiding concept has limited success outside the field of wildlife management and cannot deal with the complexity and diversity of issues associated with tourism and recreation (O'Reilly 1986; McCool 1991; Wight 1994).

Use levels

Shelby and Heberlein (1984) define carrying capacity as 'the level of use beyond which impacts exceed acceptable levels specified by evaluative standards'. It identifies a number for one management parameter: use level. It also assumes a fixed and known relationship between use level and impact parameters. So the capacity will change if other management parameters alter that relationship. Capacity will also change if management objectives change or if use populations change radically. But, as McCool (1994) indicates, little evidence exists to suggest that simply lowering or raising a specific carrying capacity standard, will bring about predictable changes in impact. It is difficult to establish a predictable linkage between use levels and impact. In addition, and importantly, the original concept deals with *process* and not with static phenomena; that is, there are natural pressures that keep populations in check – there is no one 'number'.

A problem in using use levels as management parameters is that it is all too tempting to expand limits. For example, in 1973, a Galapagos park master plan called for a cap of 12,000 visitors per year, yet by 1985 there were 17,840 visitors. Again, in 1981, a Presidential Commission report called for a cap at 25,000 visitors, yet by 1991, there were 50,000 (Bangs 1993). In spite of the range of techniques for judging and managing impact, the focus in the Galapagos has been to use carrying capacity to provide limits on numbers of visitors – limits that are apparently also ignored. The main value or strength of the carrying capacity concept may therefore be the concept itself:

> as a way of thinking about planning, carrying capacity is useful. It focuses attention on the ability of the natural environment to support growth. It suggests that development should respect the functioning of the natural processes of the environment. It shows that natural processes and man-made systems can have positive or negative effects, or both.
>
> (Schneider *et al.* 1978: 10)

The recent literature has increasingly adjusted the perspectives and definition of carrying capacity, identifying it as a management system directed towards maintenance or restoration of ecological and social conditions defined as acceptable and appropriate in area management objectives, not a system directed toward manipulation of use levels *per se*. Nevertheless, carrying capacity *does* actually refer to the maximum sustainable population (Rees 1994), whatever modifications authors have made in a *post hoc* attempt to make the concept work. In addition, experience suggests that many types of tourism stakeholders conceive of, and are looking for, 'a number'. For example, western Canada's Tourism Standards Consortium has developed certification standards for various occupations in the tourism industry. Guides are expected to understand and describe the concept of carrying capacity, which is presented as the 'maximum amount of fish, wildlife and people that can be sustained by the area over a specific time period without negatively impacting local culture, residents, environment, fish and wildlife populations and experience of clients' (Alberta Tourism Education Council 1996: 29).

There are numerous strategies and tactics for managing tourism and recreation use. Although limits certainly constitute a legitimate management tool, undue emphasis on carrying capacity may lead to narrow management policies based on rationing and use regulations (Resource Assessment Commission 1993). Practical examples of carrying capacity studies that deal with the multitude of real world variables (rather than a restricted number of isolated variables) are scarce. Carrying capacity has not provided a useful, practical tool for sustainability analysis.

Tools for sustainability analysis ────────────────────────────

Environmental impact assessment

EIA has been an important planning tool for some decades. The concept has evolved to become environmental assessment and management in recognition that it is not only a tool but also a process. EIAs identify ways of improving projects environmentally, and preventing, minimising, mitigating or compensating for adverse impacts. EIA is the major operational tool to approach sustainability at the project level currently available (Goodland *et al.* 1992).

Munro (1986) emphasised the importance of undertaking EIAs in a comprehensive policy context within a system of policy formulation, programme planning and project design and implementation. EIA should be an integral part of the planning process, and should begin when project planning begins, not after fundamental decisions, when there is only an opportunity to modify the details of the project. EIA is frequently viewed as a barrier to development – a point-in-time obstacle to be overcome for project go-ahead. This 'one-shot' approach is socially and ecologically naïve, and inappropriate for the requirements of sustainable development. Such an approach is reactive, one where the economy is considered the driving variable, and EIA is the dependent variable. As Rees (1990: 132) noted, 'what is required is a proactive approach in which the requirements of sustainability are the deriving consideration and the permissible level of economic activity is the dependent variable'.

While EIA is still largely concerned with the analysis and management of impacts on environmental and social systems caused by single-project development, over time it has become scientifically more rigorous, and its scope has gradually expanded to include:

- social, economic, cultural and other non-biophysical environmental concerns;
- programmes and policies, as well as specific projects;
- not just single projects, but also concurrent projects in an area; or
- similar projects occurring sequentially in an area; and
- cumulative effects.

In summary, however, EIA has rarely integrated environmental, social and economic issues successfully.

Cumulative effects assessment

Accumulated impacts represent 'the quantum sum of the ecological changes induced by man's use of land, water, marine and atmospheric resources' (Sadler 1986: 71). The problem of accumulating impacts on natural systems is a contemporary variant of Hardin's (1968) famous essay on 'the tragedy of the commons'. The decisions made individually (personal or corporate) are not necessarily in the best interests of an entire society. 'The cause of the tragedy [of the commons] is . . . an open or free access regime in which ownership and authority are vested nowhere' (Wackernagel and Rees 1995).

The term 'cumulative effects assessment' (CEA) carries different connotations. It can refer to the additive effects associated with a single project, (usually expected in supplementing statutory EIAs) or to the cumulative effects of multiple developments/processes, and their associated compounding effects. A definition of CEA provided by the US Council on Environmental Quality (1978) and favoured by Ross (1994: 5) is: 'an impact on the environment [that] results from the incremental impact of the action [under review] when added to other past, present, and reasonably foreseeable future actions. Cumulative impacts can result from individually minor but collectively significant actions taking place over a period of time.'

CEA is particularly important to tourism, since tourism developments not only include the large-scale resort-type projects that gain considerable publicity, but also smaller, more widespread 'mom–pop' types of operation, which may, together, have a significant impact. In addition, two of the attributes of CEA are that it is holistic and integrative (Duinker 1994).

Environmental assessment is a tool in environmental management, not an end in itself. EIA alone cannot make development sustainable, but it can help to provide the direction towards sustainability for project managers, resource managers and policy makers. It would be helpful if sustainability became a criterion for environmental assessments of projects and programmes. This would involve a move from looking mainly at the potential environmental damage of a project to how the project promotes or impairs sustainable development. The most objective and potentially measurable criterion for sustainable development is the preservation of the productivity and full functioning of the natural resource base (Mikesell 1992). Mikesell suggests that one way to apply the sustainability criterion in EIA is to measure the impact of the project on the natural resource base, and to include the negative impacts in the social costs and the positive impacts in the social benefits. All natural resource depletion and unabated environmental impacts would be charged as social costs, while any increase in renewable natural resources would be added to the social benefits. This represents an application of full environmental and resource accounting to EIA (*ibid.*).

Need for social goals and values in decisions for sustainability

Munro (1986), in describing how to improve EIAs and whether or not they involve CEA, notes that they have an important place in the chain of goal setting, strategy formulation, programme planning and project implementation, but do not themselves provide adequate guidelines for development. Munro expresses concern that EIAs, as currently undertaken, fail to take sufficient account of all factors pertinent to development decision making. While project-specific EIAs are often limited in terms of time, space or policy context, most significantly, they usually fail to relate project costs and benefits to broad social goals. Such shortcomings result largely from the fact that EIAs tend to be reactive rather than proactive. Within the context of environmental management, development must be environmentally and socially sustainable. Therefore, broad social goals are required to provide the framework within which to formulate and assess development strategies and programmes (Rhodes 1986: 32). 'These goals must take account of sustainable resource use, and continuing environmental quality, and community preferences related to the benefits of the development' (Munro 1986: 27).

Social and economic factors are the driving forces in promoting activities that cause cumulative effects. Solutions, therefore, may lie not only with improved environmental management (of which EIA is a part) but with a change in economic policies and social perceptions. As Matthews (1975) recognised, decisions resulting from EIAs incorporate subjective judgments involving values, feeling, beliefs and prejudices, as much as the results of scientific studies. The Canadian Environmental Assessment Research Council (CEARC), for example, believes that the CEA process 'can help forge a transition from project-specific environmental management to a more comprehensive 'holistic' approach to the environment' (CEARC 1988: 1). 'Values and expectations with respect to the environment, economy, and social systems, differ substantially around the world. Each particular society must define its own set of values, and each government must make policy choices based on these social ethics.'

Limits of acceptable change

LAC has been proposed as an overall framework for addressing the issues of managing impact and ensuring quality recreation experiences (Frissell and Stankey 1972). LAC is a planning procedure designed to identify preferred resource and social environmental conditions in a given recreation area and to guide the development of management techniques to achieve and protect those conditions. LAC evolved from work on design capacity in wilderness areas and on estimation of the effects of alternative use levels on the recreation environment. It emphasises identification of the objectives of a specific natural area and the development of management techniques to achieve those objectives. The procedure specifies the acceptable environmental conditions of an area, having regard to the social, economic and environmental values, its tourism potential, and other management considerations. Indicators of resource and social conditions are developed, actions for achieving acceptable standards are identified by managing authorities, and performance against standards is then regularly assessed. Condition indications relate to the state of the resource, as well as to the nature of the experience of both locals and visitors. Once desired conditions have been identified, management aims to maintain or restore those conditions. With the question focused on desired conditions and acceptable change (rather than 'how much is too much?'), the answer is more clearly one of personal judgment, not of science (Stankey et al. 1984), and echoes the need for social goals and values. This requires not only professional but also citizen input. Therefore, public involvement is a separate, but critical, component of the LAC model.

One of the advantages of LAC is that it is a forward-looking (not explanatory) process. Also, in planning visitor use, and in monitoring that use, the

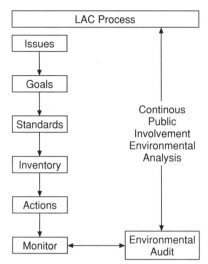

Figure 7.2   The limits of acceptable change (LAC) process.

system offers considerable benefits over the largely *ad hoc* system of development and management that exists in the tourist industry at present. It provides objective measures against which to monitor visitation/usage impacts and take remedial action.

In following the LAC process, a manager must undertake four basic tasks: (1) identify acceptable and achievable social and resource characteristics of the area being managed; (2) analyse the relationship between the existing conditions and those desired; (3) identify a series of possible management actions that will achieve the desired conditions; and (4) develop an environmental monitoring and evaluation procedure to measure the effectiveness of the management actions undertaken. Within these tasks, conventionally, LAC has nine specific steps (see Table 7.1, on page 77). However, as Wight (1994) indicated, rather than adhering strictly to the prescribed steps, it is more important to focus on the end, and to take a more general approach.

A key element in the original application of the LAC process is the definition of a series of opportunity classes. These classes describe the different conditions that the manager expects to encounter (or to restore) in different portions of the recreation area. Opportunity classes are usually designated in accordance with the divisions of the recreation opportunity spectrum (ROS). Although LAC was developed originally for wilderness management, it is a flexible process (Stankey *et al.* 1985), with potential application to other types of recreation environments, and to the full range of tourist experiences – urban, cultural and natural. What has most potential is the approach (Figure 7.2) and principles of LAC, not the specifics of the process.

The LAC system also has limitations. Unless there is detailed ecological information for each site, standards adopted will be arbitrary. In addition, in order to attract visitation to an area, some authorities may choose lower standards than are necessary to maintain the long-term environmental and cultural integrity of an area. LAC for specific locations does not appear to consider the cumulative effects of tourism-recreation activities in *surrounding* areas. Also, as practised to date, LAC does not consider whether tourism and recreation offer the wisest use of the environmental resources of an area. To tackle these drawbacks, Australia (1991) suggests:

- local and broad-based public involvement throughout the process;
- adaptation of LAC to encompass a range of resource uses, rather than the tourism-recreation sector, with more focus on the overall goals than a rigid process; and
- refinement of appropriate ecologically sustainable development indicators.

Visitor impact management

The VIM approach is also an extension of the ROS approach used to assess the impacts of visitors on a resource and recreation experience. It involves the systematic collection of data to predict the impacts of differing management strategies, and the collection of management information on the desirability of identified alternatives. VIM uses a sequential process, which aims to reduce or control the impacts that threaten the quality of outdoor recreation and tourism areas and experiences. There are eight steps in the VIM planning process (see Table 7.1, on page 77). VIM requires two separate elements: description (of the relationships between specific conditions of use and the impacts associated with these conditions) and evaluation (of the acceptability of various impacts). VIM helps to address three issues inherent in impact management (Vaske *et al.* 1994):

1 identification of problem conditions of unacceptable visitor impacts (e.g. conflicts between recreationists using a resource);
2 determination of potential causal factors affecting the occurrence and severity of unacceptable impacts; and
3 selection of potential management strategies to address the unacceptable conditions.

In addition to assisting with such issues, VIM can be integrated with other planning frameworks, or used as a management tool for a specific local impact problem. It may also be adapted for use in situations of mass tourism and in urban areas (ETB 1991).

Graefe *et al.* (1990) suggested that visitor management may be *direct* (regulate or restrict visitor activities) or *indirect* (influence visitor behaviour) (Table 7.2). Limiting use is only one of a number of strategies.

VIM was used in a consultancy to investigate the environmental and social carrying capacity of the Jenolan Caves Reserve in New South Wales, Australia (Manidis Roberts 1995). Visitation has been increasing at 5 to 6 percent a year recently, reaching over 250,000 by 1994. Despite the initial objective to determine the carrying capacity of the site, the study team, clients and advisors together determined that the focus should be on the VIM process. A team of experts in cave management and in visitor management did try to develop an

Table 7.2 Visitor management strategies

Indirect	Direct
Physical alteration	Enforcement
Information dispersal	Zoning
Economic constraints	Rationing use intensity
	Restricting activities

Sources: Cordell 1977; Hendee *et al.* 1978; Graefe *et al.* 1990.

acceptable upper limit of visitor use, but it determined that a simple number cannot be easily generated. From the environmental perspective, there were insufficient scientific data; from the social perspective, the limit depends on a range of variables, such as mode of transport, infrastructure provision and the timing of tours. The team concluded that, despite previous resource degradation, if certain visitor management activities were implemented, a 40 percent increase in the number of visitors would be possible. The project team recognised 'the application of the carrying capacity concept at Jenolan Caves Reserve may result in the over-simplification of what is a complex issue' (Manidis Roberts 1995: 9).

Visitor experience and resource protection

The United States National Park Service (NPS) has a legal obligation to address the issues of 'visitor carrying capacities' in its general management plans. However, the NPS is increasingly interpreting this concept not so much as a prescription of numbers of people, but as a prescription of desired ecological and social conditions (Hof *et al.* 1994). NPS planners have been working on developing a process to assist planners and managers to address this 'carrying capacity' requirement. The process is called Visitor Experience and Resource Protection (VERP). In a similar vein to LAC and VIM, VERP looks at the desired ecological and social conditions, rather than the numbers of people. The idea is to provide measures of the appropriate conditions, rather than measures of the maximum sustainable use.

The VERP process consists of nine steps (see Table 7.1, on page 77). In many respects, it resembles the approaches of LAC and VIM. However, VERP is intended for use throughout a park (both front-country and back-country), not simply the wilderness areas that were the traditional focus of LAC and VIM. VERP is also characterised by public involvement throughout the planning phase. The first six steps of VERP are common to the requirements of general park planning, but they also include preparing a visitor use management plan component. The last three steps require annual review and adjustment and are best accomplished as part of detailed implementation plans and management activities. The NPS optimistically indicates that in theory, every visitor use problem can be solved through design or management (*ibid.* 1994).

Like LAC and VIM, VERP also takes the approach that management goals must be translated into measurable management objectives by using indicators and standards. In addition, park zoning should reflect the management goals for different areas (e.g. developed; semi-primitive, pedestrian; semi-primitive, trekker; semi-primitive, motorised; primitive). Specific VERP indicators and standards are then developed for each of the zones: biological physical indicators, and social indicators. The latter 'measure impacts on park visitors that are caused by interactions with other visitors or with park or concession employees' (*ibid*. 1994).

This leads to a problem of using VERP as a sustainability tool beyond the objectives of NPS, in that it cannot be applied outside park boundaries if it only examines the *impact on visitors* to an area; examining the impact *of visitors* on host populations is at least equally important. VERP practitioners have correctly noted that visitor use patterns, desired visitor experiences, natural resources and park management all change with time. Thus, long-term monitoring is an essential component.

Key requirements for sustainability tools

There are a number of similarities in the approaches taken by LAC, VIM and VERP (see Table 7.1, on page 77). The review of their key characteristics reveals a number of elements that should be built into tools that are being utilised to develop sustainable tourism practices. These include ESD, public involvement, sectoral and intersectoral conflict resolution, integrated resource management, social values and ethics, indicators, scale, and monitoring.

Ecologically sustainable development (ESD)

Ecological integrity is only one of many factors that are necessary for sustainable development, particularly if it is to be fair and equitable. However, unlike some of the other factors, ecological integrity is absolutely necessary from the outset (Rees 1990). Sustainability is more than a biophysical concept; it also involves socio-economic equity (both inter- and intra-generational) and well-being in body and mind. Understanding the link between social and environmental indicators is vital. Australia (1992: 8) summarised the following principles of ESD:

- decision making should integrate long- and short-term economic, environmental, social and equity considerations;
- where serious or irreversible environmental damage is possible, lack of full information should not postpone preventive measures;
- the global dimension of impact by actions and policies should be recognised and considered;
- need to develop economic strength and diversification, which can enhance environmental protection;
- need to maintain and enhance international competitiveness in an environmentally sound manner;

- cost-effective and flexible policy instruments should be adopted; and
- community involvement is required on decisions and actions that affect them.

Public involvement

There has been a move away from narrow physical planning and promotion for tourism to a broader, more balanced approach. This recognises the needs and views of tourists and developers, and of the local and wider community. As the successor document to the World Conservation Strategy states, 'for community-based environmental management to succeed, environmental, social and economic objectives must be integrated and pursued with the full participation of the affected groups and individuals' (World Conservation Union *et al.* 1990: 31). Tourism developers and professionals should take account of local community attitudes and feelings, including the way that a local unaltered environment contributes to a community's sense of place. Participation by residents in tourism planning is fundamental. It prevents narrow special-interest groups from dictating the development process. Thus a transparent, consensus-oriented approach, with public participation throughout and continuing after development, is required.

Many techniques are available for public participation, ranging from public information to information feedback, consultation, extended involvement and joint planning. The challenge is to obtain *meaningful* public involvement on technically complex issues, as well as the effective representation of non-local perspectives. In a typical planning process, stakeholders are consulted minimally, near the end of the process, and often via formal public meetings. 'The plan that results under these conditions tends to be a prescriptive statement by the professionals rather than an agreement among the various parties'. By contrast, an interactive style 'assumes that better decisions result from open, participative processes' (Lang 1988). Planners and stakeholders seek common causes and look for solutions that are both acceptable and capable of being implemented. This approach takes a collaborative approach to problem solving, and conflict resolution.

Sectoral and intersectoral conflict resolution

One of the problems that Crowe (1969) identified was that groups who share value positions tend to make demands on that basis. Rather than use their value positions as points for bargaining and compromise with the 'opposition,' groups present them as points, which are 'not negotiable'. Crowe feels that they consciously set the stage for confrontation or surrender. That is to say, the move from scientific facts to values (whether held or assigned) increases the potential for conflict. At the same time, without conflict, major value issues are not raised and examined. Conflict is an essential part of the group process – not something to be avoided (Krawetz *et al.* 1987).

Table 7.3 Key elements of integrated resource management

- consideration of all values (ecological, economic, cultural, social and others) associated with a resource or its uses, and the effects of uses on those values in decision making
- integration of the effects of sectoral management activities within government
- integration of the effects of management between spheres of government
- integration between governments and community and industry groups

Source: Resource Assessment Commission 1993.

Integrated resource management

Integrated resource management is a fundamental precept for sustainable development (Table 7.3). There is a need to look at tourism development within the wider spectrum of potential human activities in an area. Integrated resource management does not mean multiple resource use. The latter has the implication that somehow an appropriate balance can be struck between competing uses to serve all of society's diverse demands. The result of multiple demands may simply be multiple conflicts. Instead, priority uses need to be designated. Integrated resource management requires recognition of the regional dimension of use and the spectrum of opportunity that a region can supply, as well as a flexible approach. For example, in Alberta, vast areas of public lands are designated 'multiple use'. But policy decisions have already resulted in major extractive industries being given management authority (e.g. in forest management agreements). Thus there is a problem in applying sustainability tools effectively, since conflict between sectors and activities is far more likely than collaboration (Wight 1995a).

Social values and ethics

Whatever the approach, decision makers need explicit criteria upon which to base judgments, since judgment is rarely based on scientific fact alone. When the conflict to be resolved is essentially one of values, biological information is unlikely to resolve the conflict. As a society, we must define our ecosystem ethics. Such ethics should consider the consequences of action and may vary from culture to culture or place to place. As values change, public expectations for resource management will also change. Indicators are needed to aid that judgment, since we use ecosystems in ways that allocate resources to human consumption under numerous objectives and constraints. *If quality of life is an important consideration, the problem is one of debating alternative value systems and cannot be evaded.*

Indicators

Whether indicators of sustainable development emerge from social, ecological or economic theory, the way the decision process incorporates them can influence policy and, in some instances, indicator design (Ruitenbeek 1991). Indicators are only useful in the context of appropriately framed questions. In choosing indicators, one must have a clear understanding of management goals. LAC, VIM or VERP may help in setting management goals. A typology of indicators might include:

- environmental and social indicators (measuring changes in the state of the environment and society);
- sustainability indicators (measuring distance between the present and a sustainable state of the environment); and
- sustainable development indicators (measuring progress to the broader goal of SD in a national context).

There has been a tendency to pick indicators that are easiest to measure and reflect the most visible changes, thus important issues may be dropped. Indicators to reflect desired conditions and use should ideally:

- be directly observable;
- be relatively easy to measure;
- reflect understanding that some change is normal, and be sensitive to changing use conditions;
- reflect appropriate scales (spatial and temporal);
- have ecological, not institutional or administrative, boundaries;
- encompass relevant structural, functional and compositional attributes of the ecosystem;
- include social, cultural, economic and ecological components;
- reflect understanding of indicator function/type (e.g. baseline/reference, stress, impact, management, diagnostic);
- relate to the objectives for the area; and
- be amenable to management.

Scale

Determining the levels of resolution (spatial and temporal) is a challenge for many tools and approaches. Scales must be appropriate to the questions being asked, or the zone of concern. The level of spatial resolution suitable for a detailed understanding of a small area may require considerable input data (high spatial resolution), but the level of data predictability will be low because of the limited area. For policy formulation over a larger area, there may be a higher level of predictability, but less data at this scale (low spatial resolution). This is why the preservation of large representative ecosystems and the maintenance of biodiversity have been threatened by developments being considered on the basis of narrow local issues. It is ironic that while project-level EIAs focus mainly on site scales, CEAs are designed to assess larger, regional, settings. LAC, VIM or VERP may assist in linking these two traditional scales.

Temporal scale is also an issue. Potvin (1991) presents two temporal approaches to ecosystem management. One is a static analysis (with ecosystems in steady-state equilibrium), where snapshots are compared with what is considered a desirable steady state. The other approach is a dynamic analysis (with several snapshots of a moving ecosystem), where policy aims to understand the dynamic disequilibrium transformations between the snapshots. This latter dynamic approach is more realistic, since ecosystems are not static, but dynamic, self-organising and creative. As Kay and Schneider (1994) indicated, we need to stop managing ecosystems for some fixed state. Ecosystems develop and age over time.

The concept of sustainability itself does not imply a static relationship between visitor, host community and environment. Achieving harmony is dependent on the ability to absorb and adapt to change, whether expected or unexpected. Thus, policy frameworks should be fundamentally adaptive. At present, however, monitoring agencies tend to assume that if all goes well initially, activities are within the ecosystems's capabilities. But biophysical constraints on economic policy cannot be fixed conclusively. Constraints require periodic review; monitoring is therefore intrinsic to any sustainability tool.

Monitoring

The EIA process, in part, establishes mitigation and management systems to control adverse impacts. If these systems are to be effective, then the monitoring programme must be effective. Monitoring is also important to deal with unexpected changes, which assists in adaptation and reflects the degree of resilience of the system. There is a need to be clear about what to do with the monitoring data, once collected. Data should be compatible with agency information management systems. Wight (1994) and Thompson and Wilson (1994) suggest that the monitoring system should be related to an EIA audit. Australia (1991) has also proposed that a post-development environmental audit programme be introduced as part of the EIA and development approval process. The purpose of indicators in LAC, VIM and VERP is to identify variables that will effectively indicate whether or not a desired condition is being met.

Conclusions

Simply theorising about sustainable tourism is not sufficient. There is a need not only to exhort to action, but also to contribute some practical tools towards realising tourism in a sustainable development context. In the past, determining carrying capacity has been advocated vociferously. However, in practice, this concept has very limited application. Also, while reducing use is certainly an appropriate strategy for certain problems or situations, carrying capacity is only one of many tactics within this strategy. Other strategies are also available, each with their own tactics. Other tools, or approaches, such as EIA, LAC, VIM and VERP, offer more practical and flexible application. They do not focus only on managing the use, but also on managing the resource, managing the visitor and managing the impact. LAC recognises the diversity of visitor expectations and preferences. LAC, developed for wilderness and dispersed recreation settings, has been modified for wider application, but needs further refinement to apply to more intensely used areas. VERP is a useful approach to parks management, and is intended for both front- and back-country application. However, it requires modification if it is to be used outside parks settings. VIM is a flexible process, suitable for smaller scales than LAC, and adaptable to wilderness, rural or densely populated areas. CEA is useful for extending the conventional EIA from project level to wider scales.

EIA incorporating CEA is an element of ongoing environmental management and should follow projects through their entire life cycle. Similarly, LAC, VIM and VERP should not be 'once only' tools. Common to all approaches is ongoing public participation in management and monitoring processes. All of these tools have evolved over time, and are likely to continue to do so. In the attempt to work towards sustainable development, it is the principles and the approaches, rather than the specifics, that have the most valuable application.

When different sectors have an equal or greater interest in the land, there is a need to incorporate their values; LAC assists this. VIM and VERP focus more on the causes underlying visitor impacts in a wider variety of contexts than LAC. LAC, VIM and VERP are better at incorporating human (social and economic dimensions) than ecological perspectives. A combined effort is needed among social scientists, economists and ecologists to develop tools and indicators further to accommodate a range of cross-disciplinary indicators and a variety of scales in order to help integrated analysis.

Sustainability is a dynamic phenomenon. For each tool it is necessary to consider:

- its appropriate use;
- the appropriate user;
- the implications of using the tool;
- its strengths and weaknesses;
- how successful it has been in practice;
- what other tools it combines well with;
- its ability to deal with surprise; and
- the institutional or decision-making framework required.

None of the tools presented should be seen as a panacea for tourism and resource management problems. However, they do provide valuable frameworks within which decisions can be made about acceptable conditions, priorities and resource management in a regional or more specific context.

Chapter 8

The Asia–Pacific ecotourism industry: putting sustainable tourism into practice

Alan A. Lew

According to the World Tourism Organisation (WTO), tourism in the East Asia and Pacific region grew in 1996 by 7.9 percent (arrivals) and 13 percent (receipts) over 1995 (WTO 1997a), despite problems with increasing regional air traffic congestion. These rates of growth are nearly twice that for the world overall in 1996 (4.5 percent arrivals, 7.6 percent receipts). In addition, these growth rates are expected to continue until at least 2020 (WTO 1997b). Unfortunately, the tremendous growth in tourism, while welcome economic news to many, has not been without its social and environmental costs. According to one survey of member countries of the Asia Pacific Economic Cooperation (APEC), environmental pollution, air traffic congestion and overcrowding at major attractions were the three leading constraints to expanded tourism in the region (Muqbil 1996). Private sector respondents to the same survey, however, identified excessive controls over the use of sensitive natural areas and conflicts between tour operators and natural resource managers as their major difficulties. Tension between the exploitation and conservation of tourism resources, both cultural and environmental, is acute throughout the Asia– Pacific region.

It is within this environment that the notion of ecotourism has emerged as a visible and potentially major segment of the tourism industry (Wight 1996a). Ecotourism is a very attractive and compelling concept. It reflects a growing global concern for the environment and the quality of life of communities. At the same time, however, it is very difficult to define. Definitions of ecotourism have largely reflected the many different values of speciality travel and niche market tourism (Wight 1993b; Buckley 1994). The more common of these definitions have developed out of more traditional notions of nature tourism, culture tourism and adventure travel. What ecotourism has added to these is a concern by tour operators for the well-being of the environment and local cultures where their tours take place. How this is done varies considerably, with some tour operators taking more activist roles in promoting these values, while others are more restrained in their approach. The more activist ecotour companies exhibit strong support for environmental and social causes. Companies employing the restrained approach may appear to differ little from traditional forms of speciality travel and are sometimes accused of appropriating the term 'ecotourism' for purely financial gain.

Agreeing upon a precise definition of ecotourism has, therefore, been difficult and sometimes even contentious. For the purposes of the study presented here, ecotourism is defined from the perspective of the tour provider as:

> Adventure Travel, Nature Travel, Cultural Tourism, Alternative Tourism, and Sustainable Tourism/Development that is characterised by one or more of the following: (1) a strong concern by the tour operator for the environment and local cultures where the tours are operated, (2) a company policy of contributing to local environmental and social causes where the tours are operated, and (3) the education of both guides and tourists/clients to be more sensitive in the destination environment.

This definition was adopted to assess the views and characteristics of ecotourism providers of tours to the Asia–Pacific region. It builds upon the foundations of culture tourism and nature tourism and helps to relate the present study to an earlier survey of ecotourism operators in North America by the Pacific Asia Travel Association (Yee 1992). Although its origins are different, ecotourism in its more selective and idealistic forms is recognised as a means of achieving sustainable development, itself a very idealistic concept. Many tour operators who abide by the precepts of ecotourism proponents, such as those of the Ecotourism Society, believe that they are providing a sustainable tourism product that is both environmentally and culturally sensitive and contributes to the well-being of the destinations they visit.

The focus of this study is that part of the tourism industry that has identified itself as providing an ecotourism product. These are the people and companies that put the more abstract ideas of ecotourism into practice in creating experiences for tourists. Despite the considerable literature defining what ecotourism is or should be, it is the practitioners who make ecotourism a tangible experience for their clients and destinations. In putting ecotourism into practice, they ensure that the often overlooked element of a bottom line profit margin is calculated into the definition of ecotourism. The views of the industry, therefore, are vital to understanding how ecotourism is understood and practised in the real world.

By focusing on the ecotourism industry, we are also able to avoid the much more difficult task of identifying an ecotourist. An ecotourist is assumed to be anyone who purchases a product from an ecotour operator. For many, they may be an ecotourist only during the time that they are on an ecotour. Their travel and behaviour modes may be highly unsustainable when they are not on the ecotour. Indeed, destinations and attractions are more likely to shape tourist behaviour than vice versa.

Tour operator and client characteristics

The difficulty in defining ecotourism has been a major reason that data on the growth of the ecotourism market is so difficult to obtain and to trust. This

Table 8.1   Respondent country

	No.	%
USA	22	50.0
Australia	10	22.7
Indonesia	4	9.1
New Zealand	2	4.5
Other	6	13.6
Total	44	100

Other: Canada, Greece, India, Philippines, Thailand and Western Samoa – 1 (2.3%) each.

should be kept in mind when the results of this present survey are compared with those from other sources. It also accounts for some variability in survey results, as each study's respondents will bring a different set of preconceived notions of what ecotourism is or should be. For this current study, a survey was sent to 228 tour companies that operate tours in the Asia–Pacific region. The list was obtained from the *Specialty Travel Index* (1994), participants at the 7th PATA Adventure Travel and Ecotourism Conference and Mart (15– 18 January 1995, Balikpapan, Indonesia), and the directory of members of the Ecotourism Society (1996). Seventeen of the companies were no longer in business, resulting in a total of 211 companies being surveyed. Of these, a total of 44 respondents (21 percent) returned the surveys.

A 1992 study by PATA (Yee) presented the results from 24 respondents representing North American tour companies. Of the 44 respondents to this present survey, just over half (52.3 percent) are from North America, while all but one of the remaining 19 are from the Asia–Pacific region (Table 8.1). The majority were from the USA, followed by Australia/New Zealand, and Southeast and South Asia. One respondent was from Europe. Three of the respondents were ecotourism-related consultants and developers who did not operate tour companies (Table 8.2) and were only included in tabulations that were relevant to their expertise.

Most of the respondents (84.1 percent) created and sold their own tours. This is a distinguishing characteristic of ecotour providers, which reflects their speciality tour niche market. It also allows them to respond quickly to market changes and to adjust readily to changes in a destination's accessibility and product offerings. These characteristics are as important to North America-based tour companies as to those in the Asia–Pacific region.

The international clienteles for these tour companies come from every corner of the world. Individual companies, from which the percentages in Table 8.3 (page 96) are derived, vary considerably from one another, but general patterns can be discerned. For North American companies, clients are more likely to come from the west coast (Pacific) than from the east coast (Atlantic). California, in particular, is a major source region for ecotour clients.

Table 8.2 Type of company

	No.	**%**
Buyer/reseller of tours Agent who buys tours from others to sell to the public	4	9.1
Seller/supplier of tours Creates and conducts own tours	21	47.7
Both buyer and supplier	16	36.4
Other Consultant, education – these are excluded from many of the subsequent tables	3	6.8
Total	44	100

Among Asia–Pacific-based ecotour companies, the domestic or intra-regional market is more important than the North American or European markets. This was especially true of Australia, New Zealand and Indonesia-based companies. Most Asia–Pacific-based operators also rely on the USA and Europe as important client markets.

Country and region of tour operation

Companies were asked to list all of the Asia–Pacific countries in which they conduct eco-sensitive tours (Table 8.4, on page 97). Indonesia was clearly the dominant ecotour destination overall, and scored high among Asia–Pacific and North American ecotour companies in a region-by-region analysis. Besides Indonesia, the countries of India, Nepal and Bhutan were most frequently cited by North American companies (over 40 percent each), while Australia, New Zealand and Thailand were cited most by Asia–Pacific companies.

Many secondary ecotour destinations are also shown in Table 8.4. Some of these, like China and Thailand, are already major tourist destinations, although their role as ecotourist destinations may still be developing (see Lew and Yu 1995). Other countries, such as those in Indochina and the former USSR, are emerging destinations. The countries mentioned the least are primarily small and isolated islands. Tibet, which is a region administered by China, is listed separately in this table because of its distinct identity as a destination among respondents to the survey.

The frequency of trips to individual countries was consistently higher for North American tour companies, because they tend to offer a wider array of destination choices to their clients. Asia–Pacific companies are more likely to focus on a limited selection of countries. Most of the companies that provide tour services to Australia and New Zealand, in particular, focus highly on these two countries (Table 8.5, on page 97). For example, ten of the twelve companies (67 percent) that run tours in Australia either focus on that country

Table 8.3 Major customer markets (tour company market share from foreign countries and regions for ecotours to the Asia-Pacific region.)

Companies based in North America (USA and Canada)

Client origination	Percentage of clients cited by respondents
USA	85, 85, 90, 90, 90, 90, 95, 95
California	15, 35, 45, 50
Midwest	30
New York	10, 18
Washington	6
Outside USA	2, 5, 5, 10, 10, 10, 15
Canada	5, 8, 10, 15
Australia	21, 30
Other countries cited at 2% or less: Switzerland, Argentina	

N = 12 respondents answered this question fully and correctly.

Companies based in Asia–Pacific region countries

Client origination	Percentage of clients
Asia	10
Indonesia	35, 50
Japan	15, 20, 30
Philippines	70
Australia/NZ	30
Australia	10, 15, 20, 30, 70, 80, 80
Victoria	30
NSW	30
New Zealand	15, 20
North Island	18
South Island	10
Samoa	5
North America	10, 60
USA	15, 25, 28, 50, 90, 100
Canada	20
Europe	10, 10, 20, 20, 45, 50, 50
Germany	10, 10, 35
France	40
Netherlands	25

N = 17 respondents answered most of this question correctly, although some respondents grouped countries in ways that could not be coded.

Table 8.4   Asia–Pacific countries and regions in which environmentally and culturally sensitive tours are operated (does not include tours that are not distinctly eco-sensitive)

Country or region	No. tour companies	% of all tour companies
Indonesia	16	40.0
India	13	32.5
Australia	12	30.0
Nepal	12	30.0
Bhutan	10	25.0
New Zealand	8	20.0
Tibet	8	20.0
China	7	17.5
Thailand	7	17.5
Burma	5	12.5
Cambodia	5	12.5
Laos	5	12.5
Pakistan	5	12.5
Malaysia	4	10.0
Papua New Guinea	4	10.0
Russian Far East	4	10.0
Vietnam	4	10.0
Central Asia (Uzbekistan, Kyrgistan)	3	7.5
Japan	3	7.5
Mongolia	3	7.5
Sikkim	3	7.5
Philippines	2	5.0

Countries mentioned once each (2.5% each): Easter Island (Chile), Fiji, Iran, Niue, Samoa (Western and American), Seychelles, Singapore, Sri Lanka, Tahiti, Tonga

N = 40 respondents.

Table 8.5   Number of tour companies with tours limited to only one or two countries

Country or region	No. tour companies	% of all tour companies
Asia–Pacific based		
Australia	8	20.0
Indonesia	6	15.0
New Zealand	3	7.5
Nepal	1	2.5
Philippines	1	2.5
Samoa and Niue	1	2.5
Thailand	1	2.5
North America based		
New Zealand	3	7.5
Australia	2	5.0
Mongolia	1	2.5
Papua New Guinea	1	2.5
Russian Far East	1	2.5

N = 40 respondents.

alone or combine it with just one other country (either New Zealand or Indonesia). Similarly, 75 percent of the New Zealand tour companies focus primarily on that country.

Ecotourism products

A third more nature-oriented ecotours were offered than culture-oriented tours (Table 8.6). A major difference was found between Asia–Pacific-based companies, which offer most of the 'nature' tours, and those in North America, which prefer to use the term 'adventure'. The difference here could be mostly semantic, as what might be a soft adventure tour to a North American could be a simple nature tour to a resident of an Asia–Pacific country. Trekking of one kind or another comprised the most common type of activity, followed by educational activities (Table 8.7). Water-based activities are also very popular.

Table 8.6   Types of ecotours

'Nature'
22 respondents (81.2% Asia–Pacific, 18.2% North America)
Includes: wildlife (5), nature (4), natural history (3), jungles/rainforests (2), science-based nature tours (2), fossil expeditions, national parks, nature reserves, orangutans, ornithology, village wildlife conservation zoos

'Culture'
14 respondents (57.1% Asia–Pacific, 42.9% North America)
Includes: culture (6), agriculture, anthropology, countryside tours, culture exchanges, ethnic area lodge, food, local guides, sustainable technology

'Adventure'
4 respondents (0% Asia–Pacific, 100% North America)
Includes: soft adventure (2), adventure, hard adventure, outdoor adventure

---

N = 31 respondents.

Table 8.7   Ecotour activities

'Physical – land'
15 respondents (60% Asia–Pacific, 40% North America)
Includes: trekking (7), walking (3), cycling/mountain biking (2), backpacking, bush walking, day hiking, physical activity

'Physical – water'
6 respondents (50% Asia–Pacific, 50% North America)
Includes: boat rides, diving, rafting, sailing, sea kayaking, whitewater

'Education/other'
11 respondents (42.9% Asia–Pacific, 57.1% North America)
Includes: educational (3), guest scholars/teachers/experts (3), animal riding safaris (2), birdwatching (2), local educational programmes, photo-taking safaris, study tours

---

N = 31 respondents.

This pattern corresponds with the earlier PATA survey (Yee 1992) of North American ecotour operators, which reported a preference for 'rainforest' destinations (62 percent), followed by 'islands' (17 percent) and 'mountains' (17 percent). Differences in these activities between Asia–Pacific and North American tour operators are not large, although the latter offer fewer of the landbased activities and more educational tours.

Ecotourism policies

One of the major objectives of the ecotourism approach is to support efforts at environmental conservation and the sustainable development of local communities. The use of local guides was the most common technique that the tour companies employed to support the local economy (Table 8.8). Two-thirds of the respondents said that they used local, native guides exclusively. This was higher for North American companies, for which the use of local guides provides an added element of authenticity. In addition, about two in three tour companies provided educational programmes for local guides (65 percent) and a pre-arrival information packet to tour participants (60 percent) (this latter percentage is twice that reported in the earlier PATA survey, where only 29 percent of tour companies provided printed information packets). Many operators contributed to the local environment and communities in other ways as well. As reassuring as these percentages are, the fact that many

Table 8.8   Ecotourism management policies

	No.	%
Use guides native to visited area*	31	77.5
Have an education programme for local guides	26	65.0
Provide a pre-arrival information packet	24	60.0
Provide a percentage of tour profits to local organisations	19	47.5
Participate in local clean-up programmes	17	42.5
Pack-it-out requirements	15	37.5
Other activities to support ecotourism sustainable development**	16	40.0

N = 40 respondents.
* 67% use local guides exclusively
** Other activities:

Donations: generous donations to local charities; funds for conservation and research (2); land purchases for conservation; sponsor village folk theatre; support clinic, school and religious organisations; support local environmental groups.
Education: environmental education kits; quality environmental education; scholarships; post-trip mailings; teach adult education class in ecotourism; up to 70 pages long pre-arrival packets; support village libraries; environmental reading library.
Services: provide medical services; lobby government to protect rainforest; tree planting (2); peer exchanges; support orphanages.
Economic development: use of all reusable materials; support eco-villages; encourage eco-purchases; support local handicrafts; invest in eco-lodges; support indigenous tourism projects.

ecotour companies do not practise some of these management techniques is cause for concern. Much more could be done in training local guides and educating tour group members. In addition, only 48 percent reported providing some of their profits directly to local organisations; this is well below the 75 percent reporting this in the earlier PATA survey of North American tour companies.

North American tour companies reported higher participation rates for all of the policy items listed in Table 8.8 (page 99), except for local guide training. The exception can probably be attributed to the closer relationship that Asia–Pacific-based companies have with the guides residing in their region and a stronger awareness of their professional development needs. As for the other categories, it would appear that North American tour companies are more aware of, and involved in, the ecotourism discussion and debate. They are, therefore, better able to use (and possibly abuse) the concepts and ideas of ecotourism, because they are more directly participating in creating its definition through organisations, such as the Ecotourism Society, that are based in North America.

Tour operators estimated that conducting an eco-sensitive tour cost an average 11 percent more than a similar conventional tour, although in one case as much as 40 per cent of the tour price was devoted to the special needs of an ecotour (Table 8.9). Simultaneously, only 5 percent of the survey respondents said that their clients were not interested in supporting local environmental conservation and social development projects financially (Table 8.10), suggesting that this extra cost is appropriate and acceptable, if not even

Table 8.9 Extra cost of conducting eco-sensitive tours

	% of tour price
High:	40.0
Mean:	11.1
Low:	0.0

Standard deviation: 10.2.
N = 22 respondents – close to half did not respond to this question; some stated that they did not know.

Table 8.10 Willingness of trip participants to donate money to local environmental and social causes

	No.	%
Very willing	14	38.9
Somewhat willing	20	55.6
Not interested or willing	2	5.6

N = 36 respondents.

expected, by those who take ecotours. These results point to the variability of ecotourists. Most are very aware that they are on an ecotour. While many are also apparently cost-conscious, 40 percent appear to have a strong social awareness and are very interested in participating in a form of tourism that will make a meaningful difference in the host destination (see Blamey and Braithwaite 1997). If, as Table 8.10 suggests, ecotourists recognise the added value and associated price that an ecotour entails, then it would behove the industry to ensure that ecotour providers maintain adherence to the values of ecotourism and sustainable development through the products they provide. Indeed, the failure of the industry to police itself properly in this respect has caused considerable concern over the use of these concepts and whether or not ecotourism and sustainable tourism development are fundamentally valid notions to begin with (Cater and Lowman 1994).

Managing tour groups

Only a minority of tour operators do not attempt to influence tourist behaviour in some way (Table 8.11). The degree to which this is done, however, varies considerably. While many operators enforce strict behaviour codes on most of their tours, many others are more circumspect, dealing with this issue only in very sensitive environments or if it becomes a problem. This difference is made more pronounced when comparing Asia–Pacific tour operators with those of North America. The Asia–Pacific operators are more likely to enforce behavioural rules than are North American operators. North American operators feel too removed from the local situation and prefer to leave the decision to formulate and enforce behavioural rules to the tour guide. A potential problem with this non-enforcement approach is that it places more of the burden for improper tourist behaviour and its impacts on the tour guide, which may be less dependable or consistent than broader standards adopted as company policy.

Table 8.11 Management of tourist behaviour

	No.	%
We strictly enforce sensitive behaviour on our tours	18	42.9
We explain proper behaviour, but leave it up to the individual	14	33.3
We only explain proper behaviour in the most sensitive places	5	11.9
We seldom if ever direct tourists in how to behave	5	11.9

Comments:
Our travellers typically already know how to behave – We talk to individuals privately if there is a problem with their behaviour – Our policies vary based upon the destination – Our operators are responsible for establishing proper behaviour – We do not accept participants who will not behave – Policies vary depending on the place.
N = 42 respondents.

Table 8.12   Tour group size

	Smallest group	Average group	Largest group
Mean	4.5	11.4	24.7
Median	2	8	15
Range	1–22	3–60	4–125

Do you intentionally limit tour group sizes?
Yes   34 (81%)
No   8 (19%)

If yes, what is your size limit?
Mean: 14.9
Median: 14.5
Range: 6–40

As with other forms of speciality travel, ecotourism is typically characterised by small tour groups. This was evident in the survey results, in which the median tour group size was eight (Table 8.12). Half the tour companies often had groups larger than eight and half had groups smaller than this number. Most of the respondents seldom worked with tour groups of more than 15. North American tours groups were larger than Asia–Pacific-based tour groups, probably reflecting distance and economy of scale factors. A couple of tour companies catered to very large groups (up to 125 persons), which is why the median is presented in Table 8.12 instead of a highly biased mean. Although self-identified as ecotour operators, it is questionable whether tour groups of such large sizes could realistically follow the tenets of ecotourism. Based on Table 8.13, the major reasons for preferring smaller groups were:

• to reduce the negative impacts on the environment and cultures visited;
• to allow guides to provide better service through enhanced group dynamics and individual attention to client needs;
• the limited carrying capacity of accommodation, transport and destination environments; and
• to enhance the client's experience of the destination.

Ecotourism growth

Tour companies were asked to indicate the rate of ecotourism growth that they have been experiencing over the past few years, and what they projected for 1996. The variation in the rates of growth was considerable, with some very small companies experiencing up to a ninefold increase from one year to the next. Rates this high, however, are misleading, as they may reflect a change

Table 8.13 Reasons cited for limiting tour group size

Impacts
- Reasons vary with destinations, but the impacts are greater with more than 16 persons
- To reduce/lessen impact/damage (7)
- To ensure sustainable impact
- To minimise cultural concerns/impacts (3)
- To prevent negative impacts on culturally sensitive areas
- To minimise environmental impacts (3)
- To ensure privacy
- To lower impact from camping
- We will limit tour size to one person to some pristine environments to lessen environmental and animal damage

Service
- Guides are unable to have personal contact and control the situation with more than 17 persons
- More than eight is a mob
- Ease of handling/controlling smaller groups (2)
- Some private groups may exceed our maximum
- Logistics of moving too large a group in the destination region
- Manageable, yet profitable, size
- We break our larger groups into smaller groups of four to five persons each for daily activities

Capacity
- Due to the carrying capacity of the product (2)
- Safety and the ability to airlift out of national parks and mountains by helicopter if the weather turns bad
- Our maximum size depends on the itinerary
- Depending on the destination, group sizes may be limited to as few as two persons
- Based on capacity of lodges/We are able to use smaller lodges (2)
- Allows use of smaller vehicles to get to more remote places

Experience
- To render more in-depth insight and equal service to each client
- To ensure a quality experience (4)
- To enhance enjoyment of the environment and activities
- Smaller size results in a more genuine experience
- Better group rapport/dynamics (2)
- Increased opportunity to interact with locals/cross-cultural experience (2)
- Provide more personal contact/attention (3)

---

from ten clients one year to 90 the next. By comparison, other tour companies measured their changes in many hundreds of clients per year. Consequently, the 'median' growth rate is again used in Table 8.14 (overleaf) instead of the mean.

Overall, both the number of clients taking ecotours and the revenues that they generate have grown admirably in recent years, averaging 23.4 percent and 18.3 percent, respectively, over the period from 1993 to 1995. Both were expected to increase at higher rates in 1996. The range of growth among the

Table 8.14   Ecotourism growth

	1993	1994	1995	1996*
No. of clients over previous year				
Median	68	80	107	140
Range	5–2750	0–3000	10–2500	10–2800
N	16	17	20	19
% Client growth over previous year				
Median	**	14.6	32.1	25.0
High		300	900	100
Low		−100	−25	−100
N		16	16	20
% Revenue change from previous year				
Median	10.0	20.0	25.0	29.0
High	50	50	120	70
Low	0	0	−10	10
N	14	15	17	16

* 1996 is projected.
** Insufficient data provided.

different tour companies, however, is considerable. Some have had very poor years in comparison with others. In addition, those on the original mailing list for this survey that are no longer in business are not reflected in these figures. If they were included, they would bring overall growth rates down. While ecotourism is clearly a growth industry, it is also highly competitive and not all operators have realised its popularity consistently.

The data on tour group size also raise the issue of how important ecotourism is to the travel and tourism industry overall. Ecotourism began as a speciality travel niche market and despite considerable growth and media prominence, it appears to have maintained that status. Ecotour clients tend to be highly educated and of higher than average incomes (Wight 1996a). These client characteristics allow ecotour products to exist, despite their higher cost and lower head count. Lower profit margins would require larger group sizes to maintain a tour company's economic viability. This would lower the perceived value of the product as a speciality tour, while raising concerns about the environmental and social impacts of tourists in ecologically sensitive areas. It would appear that for ecotourism to maintain its viability as a type of tour experience, it will need to remain a speciality tour product. While the values that ecotourism expresses may have some resonance in the travel industry overall, the economic impact of the ecotourism niche market is likely to remain small.

Ecotourism industry trends

The survey contained three questions that addressed trends in the Asia–Pacific ecotourism industry. The first question asked how the industry has changed over the past two years (between about 1993 and 1996). The second question addressed projected trends over the coming two years, while the third asked respondents to indicate barriers to the future development of ecotourism. The most pronounced trend of the past two years has been a broadening of the clientele for ecotours and an increase in FITs (independent, non-group travellers). More people of varied ages and incomes are taking ecotours, leading to a softening of the adventure element and greater sensitivity to pricing. Also, these new clients are more aware of the environmental and social development issues of the destinations they are visiting.

For the future, the responding tour companies see a broadening of market channels through both the Internet and better global communication systems. Ecotourism will continue to expand into new areas, such as Central Asia, and will offer new products, such as day tours for FITs. The respondents also predict an increasing awareness of ecotourism and eco-friendly tourism overall. While these trends appear to show that ecotourism is about to break out of the limited speciality travel market, they actually reflect broader patterns found throughout the tourism industry. Increasing affluence worldwide has helped to fuel strong growth in all segments of the tourism market, including speciality travel. What may be more important in the long term is the gradually increasing awareness of the need to introduce more sustainable development values in the products and practices of the tourism industry.

Many problems and barriers to the development of ecotourism were also identified. The first is a general lack of knowledge about ecotourism in the travel industry, among most tourists, and within society in general. This is reflected in the political arena in a mistrust of ecotour operators by government officials. The tourism industry itself presents problems in the lack of standards for quality and pressure to expand mass tourism into this speciality niche market. Several respondents felt that growth projections for ecotourism have been too high, resulting in too many tour operators. High costs due to a lack of infrastructure were cited by some as a problem. However, as more remote places become better connected to the outside world, these prices are likely to fall, along with the loss of natural environments and traditional cultures that form the basis of many ecotour experiences. This is the price of the modernisation that most countries seek.

Conclusions

The 1992 PATA ecotourism survey (Yee) identified three general issues facing ecotourism: (1) standards, (2) education, and (3) tourism policy and carrying capacity issues. Five years later, these issues remain as major concerns for the industry. International organisations such as the Ecotourism Society have made great strides in addressing the issues of standards and education. However, the world is a big place and reaching the grassroots-level tour operator is difficult. Both governments and the larger, mainstream tourism industry are at least giving rhetorical voice to the values of ecotourism and sustainable development, although in practice these policies are seldom carried out on a large scale.

The results of this survey suggest that ecotourism remains a small, though growing, industry. Most of the practitioners of ecotourism have considerable faith in the work that they are doing and are trying to be responsible to the local environments that they tour. They face major problems and challenges in achieving their vision of a sustainable form of tourism. The biggest barrier is that because most of their success have been based on inherent characteristics of their speciality travel niche market, they cannot be recreated within the larger mass tourism and travel industry. Some broader social trends, however, are in their favour, including the development of improved communications and marketing systems (e.g. the World Wide Web) and a growing eco-awareness among people worldwide. By contributing to these trends, ecotourism is helping to bring about a more sustainable and ecologically sensitive tourism industry.

Differences can be readily seen between ecotour operators based within the Asia–Pacific region and those based in North America. Simple location and accessibility to Asia–Pacific destinations account for some of these differences. It would be expected, for example, that companies that specialise in taking groups overseas would carry a similar broad mix of products, as do the North American operators that take groups to the Asia–Pacific. Likewise, domestic tour companies in any region of the world would focus more on the products offered in their home country than overseas operators. Other differences may be partly semantic in that different cultures define and use words in different ways. Such differences, however, present real problems for tour operators and tour participants in a market as highly specialised and increasingly competitive as ecotourism is. As the travel industry comes to understand these differences, a greater consensus should emerge on what ecotourism is and what it is not.

Chapter 9

How sustainable is ecotourism in Costa Rica?

Susan E. Place

Nature-based travel, now generally called ecotourism, began to take off in the 1980s and has since become one of the fastest-growing segments of the global tourist industry. Ecotourism represents a response to phenomena occurring in both centre and periphery, deriving from the dominant global economic paradigm based on continuous growth. In the centre, industrial and urban development caused destruction of natural ecosystems and degradation of the environment in general. At the same time, increasing affluence led to growing demand for natural areas for recreation and travel (Nicoara 1992). Affluent people from developed countries increasingly sought out exotic wild places in Latin America, Africa and Asia, where less space had been fully integrated into the global economy. In the periphery, stagnant or declining commodity prices and growing foreign debts forced governments to seek economic alternatives. Pressured by multilateral institutions to restructure their economies, many countries sought to diversify production, particularly in terms of exports. Some countries, such as Costa Rica, decided to promote tourism as a foreign exchange earner. Banking on its world-renowned national park system, Costa Rica aimed for the niche market of ecotourism, which meant that it had to promote conservation of its remaining natural areas. During the past decade, in Costa Rica and other third world countries, protected wild areas have been increasingly integrated into the global economy through ecotourism.

Environmentalists have promoted tourism as a non-consumptive use of nature and a win–win development strategy for underdeveloped rural areas. As an influential World Wildlife Fund publication on ecotourism states:

> One alternative proposed as a means to link economic incentives with natural resources preservation is the promotion of nature tourism. With increased tourism to parks and reserves, which are often located in rural areas, the populations surrounding the protected areas can find employment through small-scale tourism enterprises. Greater levels of nature tourism can also have a substantial economic multiplier effect for the rest of the country. Therefore, tourism to protected areas demonstrates the value of natural resources to tourists, rural populations, park managers, government officials and tour operators.
>
> (Boo 1990: 3)

107

Analysis of ecotourism in countries such as Costa Rica will reveal whether or not it, in fact, generates the benefits described above. Local and national economic benefits of foreign tourism need to be examined in the context of its potential costs. In the case of ecotourism, the elimination of competing land uses is usually required. Farming, forestry, mining and sometimes hunting are excluded from most protected wildlands. Income from these activities must be replaced by tourist-generated income. Thus we must ask where most foreign tourist revenues go. Most never make it to the destinations favoured by ecotourists. A high proportion of the tourist dollar goes to outsiders such as hotel and tour operators (often foreign), airlines and foreign travel agents. As a result of this economic leakage, natural resources that once provided livelihoods for local people now generate profits for outsiders, either in the principal cities of the destination country or in the source country of the tourists (Wallace 1992; Honey 1994). Only a relatively small proportion of the tourist dollar remains in the actual destination, largely in the form of wages to employees in tourist services. Furthermore, while the benefits of ecotourism accrue to affluent national and foreign entrepreneurs, its costs may be borne disproportionately by the rural poor. Not only do local people lose their resource base for farming, lumbering or mining, but they may also lose the subsidy from nature upon which their livelihood was based (Hecht *et al.* 1988). In other words, they may lose access to forests that provided them with fuel, construction materials, wild foods and medicines at little cost other than labour and time. They may also lose access to beaches where they fished, swam and engaged in various leisure activities. In addition, the local populations will feel the most sustained impact of the congestion and pollution generated by tourist facilities.

The provision of tourist infrastructure may compete with the people's basic social needs for scarce public funds. The imperative of providing improved services for tourists may prompt governments to invest in airport construction, highway and port improvements, and costly provision of infrastructure such as electricity and sewage treatment to remote locations (where ecotourist destinations are usually located). At the same time, government funding of basic social services such as health care, education and food and nutrition programmes, may stagnate or shrink, even as population and need increase (Howard 1993; editors of *Rumbo* 1993).

Cultural degradation may also occur, as formerly isolated populations come into contact with affluent foreigners (e.g. Wilkinson 1989; Boo 1990; Hill 1990). Ecotourism probably represents less of a threat than mass tourism, at least in terms of problems such as prostitution. Nonetheless, local populations may modify traditional folkways, from diet and handicrafts to song and dance, to meet the different cultural standards and expectations of foreign tourists.

The role of park-based tourism in Costa Rica

Costa Rica is an appropriate place to examine the role that protected wildlands and ecotourism can play in economic development, and the contradictions that emerge in ecotourist destinations. During the past decade, Costa Rica has aggressively pursued the promotion of tourism as the centrepiece of its development strategy. By the time the Oscar Arias administration (1986–90) had decided to make tourism Costa Rica's number one foreign exchange earner, the country was already well known among naturalists and ecologists for its natural beauty, biodiversity and existing system of protected wildlands (Laarman and Perdue 1989). Thus the government, quite logically, decided to promote nature tourism (using the advertising slogan of 'Costa Rica: naturally'). The success of this campaign is demonstrated by the fact that 75 percent of international tourists visit Costa Rica for its natural areas (Mora 1996).

The cessation of overt military action in neighbouring Central American countries, as well as the publicity surrounding Oscar Arias' Nobel Peace Prize, led to a rapid increase in foreign tourism to Costa Rica. The number of tourists rose from 376,000 in 1989 to over 610,000 by 1992 (Leininger 1993). By the mid-1990s, tourism had become Costa Rica's leading source of foreign exchange (Mora 1996). The national parks also experienced a rapid increase in foreign tourists, with the number more than doubling between 1988 and 1992, after already having doubled between 1984 and 1988 (Place 1995).

Costa Rica has a significant amount of domestic tourism, unlike many other third world tourist meccas, which is probably partly a reflection of the substantial size and high educational level of its middle class (Hill 1990). The national parks have become an important source of national pride and a focus of domestic tourism, reflecting a growing interest in environmental issues and conservation of the country's unique biological endowments. The number of Costa Ricans visiting the national parks increased from about 130,000 in 1982 to over 300,000 in 1992. In the 1980s, Costa Ricans of modest income were able to reach some national parks on public transport, and once there, could stay in inexpensive lodgings within the means of even working-class families. This is important, because domestic tourism can help to mitigate the effects of volatility in the flow of international tourism. Later in the decade, however, as the Costa Rican tourist industry shifted its focus towards more lucrative foreign travellers, the cost of accommodation near parks and reserves skyrocketed, so that even middle-class Costa Ricans were being priced out of vacations in their own country, which was a source of considerable resentment by the early 1990s.

The question of who benefits from tourism fits into a long-running debate over the role of tourism in third world development (e.g. Bryden 1973; deKadt 1979; Murphy 1985; Smith and Eadington 1992; Mora 1996). Certainly, in Costa Rica the recent tourist boom has increased foreign exchange receipts. But who actually receives most of the income? Investors, many of them foreign, have rushed to capitalise on the exponential growth in tourism. As a result, between 1989 and 1992 investment in tourism grew by 87 percent (Howard 1993). However, by 1996 the boom was losing its glow. Increases in tourism had not reached the levels projected three years previously and there were empty facilities (Mora 1996), highlighting the risky nature of tourism as a development strategy. Unemployment is likely to increase under such circumstances, and small, marginal tourist enterprises are the most likely to go under.

Tourists themselves may become investors through the purchase of land or second homes, which thousands of foreign visitors have done. While it is a tribute to Costa Rica's charm, it has also distorted the country's real estate market. Prices have inflated so dramatically in the past five years that many Costa Ricans have been priced out of land and home ownership (Howard 1993). Costa Ricans, once primarily independent family farmers, now find themselves increasingly relegated to providing low-waged services to foreign tourists while generating profits for foreign entrepreneurs.

Tourism and development in Tortuguero

This chapter focuses on a case study, the Caribbean village of Tortuguero, whose recent transformation reveals many of the conflicting interests in the tourism-based development process. Tortuguero's ecotourism-based economy also illustrates the linkages between local, national and international scales that underlie environmental relations and economic development in general. While Tortuguero exemplifies the potential of tourism to provide for local or regional development, it also reveals some of the dangers of this economic strategy. Tourism is a notoriously volatile economic activity and is subject to booms and busts. Tortuguero's experience shows how even remote rural areas are affected by outside events and trends, often the result of decisions made by outsiders and over which local people have no control. Like other tourist destinations Tortuguero is vulnerable to the vagaries of the world economy, the political climate and natural disasters (such as earthquakes). This section considers the question, who in Tortuguero has benefited from the tourism boom?

In 1986, the author spent eight months studying the local effects of the establishment of Tortuguero National Park in northeastern Costa Rica (Figure 9.1), a place she had visited periodically since 1975 (Place 1988). As a result of this fieldwork, she became interested in the impact of nature-based tourism on the local economy, and its potential to replace income from the exploitation of resources now protected by the park (Place 1991). In May 1993, the author returned to Tortuguero, where the changes accompanying the rapid growth of tourism were assessed (Place 1995).

Tortuguero National Park was created in 1975, primarily to protect the last remaining major green turtle nesting beach in the western Caribbean. About 20,000 hectares of lowland tropical forests adjacent to the 30 km beach were also included within the park, protecting them from destruction by the rapidly moving frontier of colonisation on Costa Rica's Caribbean coastal plain. The park is located in one of the wettest regions of Costa Rica, which receives an average of 5000 mm of rain a year.

The village of Tortuguero, located on a barrier island along the turtle nesting beach and on the northern border of the park, is isolated from the rest of Costa Rica by swamps, rivers and estuaries. There is no road to the village,

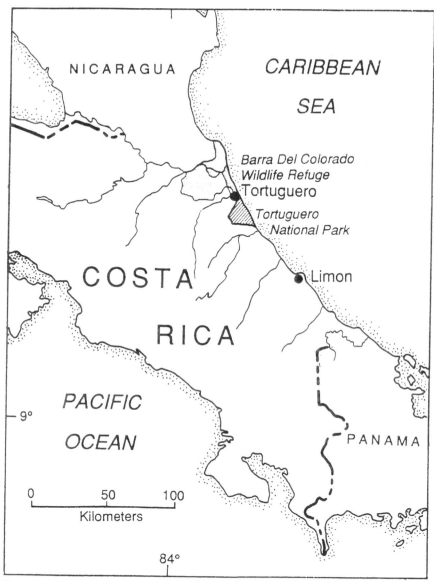

Figure 9.1   Tortuguero, Costa Rica.

and it did not receive an electricity line until the end of the 1980s. From its founding, early in this century, the economy of Tortuguero has been based on the exploitation of biological resources for sale to the international market and for subsistence. The first settlers hunted turtles and sold them in Limon, Costa Rica's major Caribbean port (about 80 km south). Turtling declined by the 1970s, both because of dwindling turtle populations and conservation measures that prohibited hunting them. Logging for the export market began in the 1940s. After a boom period in the 1950s and 1960s, logging declined and the village sawmills closed in the early 1970s. Villagers began to emigrate, seeking work elsewhere. By 1986, Tortuguero's population numbered only about 150 people, who were reduced to subsistence farming, seasonal wage labour, remittances, and returns on investments made with income from previous economic booms. An informal economy thrived within the village, so that any infusion of money cycled extensively through it and allowed villagers to survive on surprisingly low formal incomes (Place 1988, 1991).

In the 1980s, all of the formal employment in Tortuguero existed because of local biological resources, which by then were protected in the national park. Although the number of people employed by these entities was small, it represented a significant sector of the village economy. For example, 19 of 27 households received some income from services to tourist or research facilities. Most of these villagers, however, worked only part-time or seasonally. About half the villagers surveyed in 1986 felt that their standard of living had declined in the previous decade, while half of the remainder felt it was about the same. Most of the 25 percent who thought life was better were receiving economic benefits from the influx of outsiders, attracted by the park or biological research at the green turtle research station near the village. Surprisingly, even these villagers were largely unaware of the role of the national park in the incipient tourist trade. They did not appear to consciously question why tourists were beginning to arrive in larger numbers.

The national park is not the only manifestation of the importance of government policy to this remote village. In the mid-1970s, a state agency completed a canal linking Tortuguero with Limon, providing the first permanent connection between Tortuguero and the rest of Costa Rica. The canal, in conjunction with the national park, significantly altered the economy of the village. The tenuousness of this link, however, was revealed by the massive April 1991 earthquake on the Caribbean coast (7.4 on the Richter scale). The trembler caused such significant tectonic uplift that the canal became unnavigable, eliminating public transport to the village, and reinstating its former isolation. The loss impacted on local entrepreneurs because several villagers offered rustic accommodation in *cabinas* and provided typical meals in *comedores*. The owners of these businesses purchased fish, game meat and produce (when available) from local people, thus helping to spread the income from tourism. Most of their clientele usually arrived by public transport. As a result, the earthquake-induced elimination of public transport threatened the economic viability of these village entrepreneurs.

State economic policies, in addition to the provision of infrastructure such as transport and energy, also affect even remote villages like Tortuguero. The 1980s brought a series of economic crises to Costa Rica, which resulted in structural adjustment that reduced public expenditure while encouraging private investment. Under pressure from various international institutions, the Costa Rican government decided to promote foreign tourism as a way of diversifying the country's economy and reducing its trade imbalance. That policy had a profound impact on Tortuguero. The number of foreign visitors

Table 9.1 Annual number of visitors to Tortuguero National Park

Year	Total number of visitors	Domestic visitors	% of total	Foreign visitors	% of total
1980	226	n.a.		n.a.	
1981	674	378	56	296	44
1982	843	488	58	355	42
1983	n.a.	n.a.		n.a.	
1984	1,863	1,020	55	843	45
1985–86*	2,018	999	50	1,019	50
1986	2,004	972	49	1,032	51
1987	n.a.	n.a.		n.a.	
1988	2,776	1,066	38	1,710	62
1989**	3,270	429	13	2,841	87
1990	9,207	1,155	13	8,052	87
1991	15,646	389	3	15,257	97
1992	19,741	50	0	19,691	100

* Based on entries in park visitors book, 18 May 1985–17 May 1986. Data not available for calendar year 1985.
** Data appear to be incomplete. Five months show as receiving no visitors, a highly unlikely event. The total for the remaining seven months suggests a substantial increase in annual visitation to the park. (The number of domestic visitors appears to be under-reported.)
Source: Unpublished data from Servicio de Parques Nacionales.

to Tortuguero National Park began to increase exponentially in about 1990 (Table 9.1). In 1982, only 355 foreigners visited the park (about 42 percent of total visitors), while 488 Costa Ricans visited. In 1992, the park received 19,691 foreign visitors and only 50 Costa Ricans! The flood of foreigners has swamped the former village population of 150 people, and has attracted scores of immigrants to work in the new tourist enterprises, transforming the village. The question is whether this transformation has improved the lives of the original villagers, those living there in the decades before tourism arrived, who are presumably the target of rural development policies.

Local entrepreneurs have shown an interest in providing services to the growing number of tourists visiting Tortuguero, and with minimal help could contribute a substantial amount to building a tourism-based economy with a number of local economic multipliers. Ironically, the rapidity of the increase in foreign visitors to Tortuguero may be reducing the opportunity for village entrepreneurs to become involved in the tourism business other than as menial employees in dead-end jobs. The pace of outside investment is not allowing the villagers the luxury of time to accumulate their own capital to invest in tourist facilities and services.

By the mid-1980s, as tourism began to take off, outside investors were already moving in to capture the growing tourist market in Tortuguero. In 1985, two Costa Rican travel agencies began to provide weekend package cruises along the inland canal from Limon to Tortuguero, lodging clients overnight in Tortuguero, primarily using rustic *cabinas* owned by villagers. Initially, these tours were oriented towards the Costa Rican middle class, who were interested in Tortuguero's 'exotic' Afro-Caribbean culture and environment perhaps even more than in the national park's offering of wild nature. In 1990, however, these tour operators opened their own hotels just outside Tortuguero. In response to the government's successful promotion of foreign tourism, the new hotels were oriented towards the affluent international traveller. The high prices of the new hotels effectively precluded most Costa Ricans from visiting Tortuguero, at least on the package tours that had become so popular since 1985. This reveals the importance of tour operators in directing the course of tourism development and underscores the lack of power that rural people have to control their own economic development process in the face of a global juggernaut such as tourism.

In the early 1990s, another outsider constructed a large, purple, cement-block souvenir shop in the middle of the village (otherwise composed of one- or two-room thatched-roof and weathered wooden dwellings). This souvenir shop sells handicrafts and other items that are imported, and it has little relationship to the inhabitants or culture of Tortuguero village. Local people have already lost the opportunity to develop their own handicraft businesses, unless they can produce something that they can persuade the hotels to market for them or organise to create a crafts outlet to compete with the new souvenir shop. As handicrafts were not part of the traditional culture of the village, this is unlikely to happen.

Other factors beyond the control of local populations, such as natural hazards and government policy in response to them, also affect the course of economic development. As mentioned above, the April 1991 earthquake left the village isolated because part of the canal became unnavigable. By May 1993, the canal still had not been repaired. This eliminated the public transport that had linked Tortuguero to the rest of Costa Rica. The only way for tourists to reach the village was to charter a plane or pay for the pricy package tours (which included lodging at the expensive new hotels). This contributed to a dramatic decline in domestic tourism to Tortuguero, while the number of international visitors took off exponentially. By May 1995, the canal had been reopened, but a lack of regular public transport remains a problem for villagers and domestic tourists (Blake and Becher 1997), illustrating the vulnerability of remote ecotourism operations to disruptions beyond their control.

The new orientation towards foreign tourists significantly altered the village population. A visit to Tortuguero in May 1993 revealed that scores of new residents had moved into the village and its environs to take advantage of employment opportunities opened up by the new tourist facilities. At the same time, a number of the old-time residents had left the village. The traditional villagers tend to have less education and fewer of the skills necessary for employment in the new tourist businesses than do the outsiders. This calls into question the assertion that ecotourism promotes local employment.

The tourists who stay at lodges oriented towards international travellers spend a far greater amount *per capita* than those who stay in the village. Sport fishers paid over US$100 per day (in 1986) to stay at Tortuga Lodge, while tourists who stayed in the village paid only a few dollars a day. However, economic leakage is high at the expensive hotels, since most guests book their tours through agencies in San José or their home country, so little of the money paid for a visit to Tortuguero stays in the village except for hotel workers' salaries and tips to local guides. While the lodges provide a considerable amount of low-paying, part-time employment to villagers, they do not provide a reliable source of capital for investment in the village's future.

If tourism is to promote community development in Tortuguero, it ought to focus on expanding, improving and promoting locally owned facilities and services. By the mid-1990s, several locally owned lodgings had been established (Blake and Becher 1997), indicating that tourism was beginning to produce some local economic benefits. Such village-based tourism, oriented towards domestic and foreign tourists on a budget, has the potential to produce more local economic multipliers than the high-priced package tour hotels. Both women and men can participate in the economic spin-offs from such tourism. Villagers of both sexes, for example, farm and fish. Village lodgings are owned by both women and men. Men, however, are more likely to participate in the most remunerative activities, such as hunting and guiding tourists and sport fishers.

The potential for local entrepreneurs to organise for the purpose of capturing a larger share of future income from tourism is impeded by the atomistic nature of the village of Tortuguero. Rather than forming a community bonded by kinship and other social networks, Tortuguero resembles a collection of individuals occupying the same space. Most of its inhabitants migrated to the village to take advantage of opportunities offered by turtling, logging and now tourism. Thus, they tend to relate to each other more competitively than cooperatively. However, it appears that by 1996 some degree of village organisation had coalesced around local development issues (Blake and Becher 1997). Given the history of the village, it remains to be seen if cooperation will become the dominant mode of interaction in Tortuguero.

Another obstacle to village business creation is that government and financial institution policies discriminate against small local enterprises. When the Costa Rican government decided to promote tourism aggressively as a development strategy, it focused on luxury beach resorts, and other forms of mass tourism, in order to maximise capital accumulation. For example, a 1985 Tourist Development Incentives Law provided for tax breaks for businesses serving tourists. To qualify for tax incentives, however, a business had to meet stringent criteria regarding the design of the facility, including that it contain a minimum of 20 rooms with baths (Hill 1990), something a village entrepreneur could not possibly afford.

Conclusions

Although the Costa Rican government has paid lip service to nature-oriented tourism, it has offered little material help. On the contrary, among the casualties of structural adjustment were the park service, forest service and other resource conservation budgets (Carriere 1991). Despite the budget cuts, national parks such as Tortuguero will undoubtedly continue to attract scientific and nature-oriented tourism (both domestic and international) because of the uniqueness of its green turtle rookery and, increasingly, the rarity of the tropical rainforest biome (Carr n.d.). Tortuguero has the prerequisites for sustainable conservation-based tourism: an interesting and relatively easily viewed wildlife component, the potential for reasonably easy access, and a distinctive local culture, which has long been in contact with the outside world. Even so, a positive outcome from tourism is by no means assured and would seem to depend on how development is planned and implemented in Tortuguero.

If ecotourism is to form a sustainable economic foundation for northeast-ern Costa Rica, careful planning is necessary to keep the development process under control so that it does not grow too rapidly and so that some role is preserved for local entrepreneurs. Furthermore, Tortuguero National Park needs a larger budget, especially for personnel and infrastructure (especially visitor facilities), and it should be integrated with the new Barra del Colorado Wildlife Reserve to the north to create a regional circuit of natural areas. Local inhabitants should be more directly involved with the park and wildlife refuge, and with the ecotourism attracted to them. There are several ways to do this, which would require only minimal training for some villagers.

One option would be to hire local people who are knowledgeable about the area's natural history as park naturalists and river guides (after appropriate training). The park could, for example, encourage villagers to run canoe liver-ies and guide services. A few villagers are beginning to do this, but canoe rentals are still not integrated into the park (Blake and Becher 1997). Several villagers would be very well suited to such activities, especially those who speak English. Unfortunately, most do not have the capital to start their own businesses. Such enterprises could be run from within the park, providing important visitor services (without any cost to the park) and promoting the integration of village and park.

Domestic tourism to Tortuguero should be promoted to counterbalance the increase in foreign tourism. The experience of the 1980s suggests that there is a sizeable potential market among the Costa Rican population and that they are most likely to stay in modest *cabinas* and patronise locally owned establishments. Thus domestic tourism is much more likely than foreign tourism to support village-based development. The promotion of domestic tourism, however, will depend on the provision of inexpensive and reliable public trans-port to Tortuguero. Furthermore, village entrepreneurs need seed money, per-haps by means of a rotating credit bank, and small business training. There are villagers who want to and are capable of expanding tourist services such as *cabinas*, meal services, boat rental and guide services. Such local entrepreneurs ought to be enabled to take advantage of the growth in tourism before their options are completely foreclosed by outside investors.

Despite recent recognition of the close relationship between environmental health and sustainable development, for the most part there does not appear to be much understanding of the complexities of this relationship. While parks and preserves may perform important ecological functions by protecting biodiversity, watersheds and soils, they also represent the loss of vital resources to local inhabitants. The Costa Rican government may have successfully protected endangered species and environments in Tortuguero National Park, but it has not done anything to ease the local population's transition from an economy based on resource extraction to one based on the preservation of ecosystems. The stagnation of Tortuguero's economy in the decade after the creation of the national park illustrates this. If ecotourism proves capable of providing sufficient economic opportunities to residents no longer able to exploit natural resources, nature conservation programmes might be able to support sustainable rural development.

The contradictions inherent in nature-based tourism, however, prompt questions about the whole notion of sustainable development. For example, the rapid increase in foreign tourism to Tortuguero, and the capture of most of the profits from it by outsiders, exemplifies the problematic relationship between environment and development. The income generated by the 'nonconsumptive' use of local resources now flows to the outsiders who bring the tourists to Tortuguero and provide their accommodation and other services. Thus travel agencies, transport companies, tour operators, hotel owners and souvenir shop owners reap most of the benefits of tourism-based development, rather than the local inhabitants who used to depend on nearby natural resources for their livelihoods.

Ecotourism can provide an alternative economic base, but it does not happen automatically, or without social and environmental impacts. If it is to be sustainable, local populations must be allowed to capture a significant amount of the economic multipliers generated by tourism. Successful reduction of multiplier leakage requires local participation in development planning and outside assistance with the provision of necessary infrastructure, training and credit. Community participation is also essential for identifying negative impacts on people who live in areas undergoing ecotourism development. Yet there are powerful internal and external obstacles to local participation, from factionalism within communities to state policies that promote centralised planning and the accumulation of capital among large tourism enterprises.

Costa Rica, like other third world countries that have based their development strategies on tourism, faces an important junction in its path to development. The dominant development paradigm promotes the sort of dependent development that leads to the 'kidnapping of a nation', as suggested by the Costa Rican scholar Eduardo Mora Castellano. Mora is concerned about the perceived need in Costa Rica to create a fantasy experience that meets the expectations of foreign visitors (see Olwig and Olwig 1979). Since foreign investors are more likely to succeed in this approach to tourism, small local businesses are likely to be squeezed out, leaving the country in the hands of foreign image makers and profit takers. On the other hand, there is still room for an alternative route to development, based on grassroots initiatives and participatory planning that can promote more authentic, sustainable tourism.

Some communities, such as the Kuna Indians of Panama (Breslin and Chapin 1984), have succeeded in taking some control over local development and can provide inspiration and concrete ideas for the development of other areas in danger of being overtaken by ecotourism. In Costa Rica, a national eco-agricultural cooperative network, COOPRENA, has begun to promote community-based ecotourism initiatives. Six farming cooperatives are now offering ecotourism as part of a diversified livelihood strategy designed to use their land productively while also conserving natural resources and generating employment and other socio-economic benefits (Green Arrow 1997). There are many communities around the world engaged in this struggle against cooptation by the tourism industry. Systematic collection of their stories, and analysis of their successes and failures, could help to provide support for people's future quests for self-determination in community development.

Acknowledgments

The research for this chapter was made possible in part by a Fulbright-Hays Faculty Research Grant. Guy King, Beth Mills, Chuck Nelson, and D.J. O'Donnell kindly provided assistance with various phases of map production.

Chapter 10

Curry County sustainable nature-based tourism project

Bill Forbes

Curry County, located on Oregon's south coast, has attempted to promote economic diversification through sustainable nature-based tourism. Traditional tourism in Curry County has relied on jet boat trips, sport fishing and drive-through tourism along coastal US Highway 101. The county has attempted to develop more reliable, year-round markets based on its world-class nature tourism resources: scenic coastline, old-growth redwoods, and unique botanical and geological diversity.

The Siskiyou National Forest, through its rural community assistance programme, has acted in partnership with Curry County to bring in prominent professionals in sustainable tourism as advisors. The project sought to develop sustainable tourism by including (1) sustainability planning; (2) product development; (3) business training; and (4) marketing.

At the time of writing, the project was nearing the end of a four-part initiative developed by tourism advisors Egret Communications. Local agencies and citizens were taking over the project in the autumn of 1997. Seven new businesses were starting up in a variety of product development categories. Considerable interest had been generated in the possible development of an elevated forest walkway and accompanying visitor centre.

A solid sector of local support had been established. A citizens' environmental and social assessment committee was established and working. A non-profit tourism board has also been set up, comprised of a broad representation of locals with knowledge and support of the concepts of sustainable tourism.

Community profiles and natural setting

Curry County is located on the southernmost coast of Oregon. It is bordered by California to the south, the Pacific Ocean to the west, Coos County to the north and the Siskiyou Mountains to the east. Its population is approximately 22,000. The base economy relies on, in order of importance, logging, tourism, commercial fishing, agriculture (cranberries, lily bulbs) and retirees.

Settled by pioneer families during and after the Rogue Indian Wars of the 1850s, its land ownership today is largely a mix of historically family-owned holdings (5 percent), timber company land (10 percent), national forest

(55 percent), and state parks and other government holdings (10 percent). There is a strong tradition of public land use by locals for hunting and fishing. The county has only three main communities. From south to north, and in order of approximate size, they are Brookings–Harbor (10,000), Gold Beach (3,000) and Port Orford (1,500). The three hamlets of Agness, Langlois and Pistol River also have their own community identity.

The economy of the Brookings–Harbor area centres around the South Coast Lumber Company (approximately 400 employees) and its plywood mill and sawmill, commercial and sport fishing, drive-through tourism, lily bulb growing (supplying about 90 percent of the easter lilies in the USA), a new state prison 20 miles south in California, and a growing retirement community. The atmosphere of the community is one of a mill town with beginning strip development, a downtown and port that service mostly locals, and outstanding surrounding coastal and mountain scenery.

Traditional visitor traffic to the Brookings–Harbor area has centred on sport fishing and northbound traffic visiting the scenic Oregon coast. Recent fishing restrictions, due to declines in native fish populations, have brought uncertainty and reductions in the traditional tourist traffic. As part of local tourism development, a boardwalk with shopping sites is being constructed at the port with US Department of Agriculture (USDA) rural development funds. Local leaders are very pro-growth oriented, having recently approved the largest urban-growth boundary expansion in Oregon's history.

Gold Beach has centred its economy on tourism. Jet boat trips and sport fishing on the Rogue River attract many destination and drive-through visitors. Motels line coastal Highway 101 through the length of the town. Visitor revenues surpass those of the larger Brookings–Harbor area, and the town re-invests its substantial motel taxes in advertising. Despite the existing emphasis on tourism, the town could further benefit from more reliable, off-season visitors who spend money on package tours.

Port Orford is the smallest of the three main towns. It lies in a picturesque setting between coastal Humbug Mountain (457 metres) and rugged headlands to the north. Its economy relies on commercial fishing, productive cranberry bogs, and some farming and logging. It has the lowest *per capita* income of the three communities. While Port Orford lacks the infrastructure for substantially increased tourism development, it presents opportunities for small businesses and tours in its outstandingly attractive surroundings.

In comparison with other Northwest timber-dependent counties, Curry County has more economic bases (agriculture, prison, retirees, tourism) and a relatively low percentage of unemployed timber workers. The two major mills in Brookings are still operating. Most of the lumber mills in the north end of the county began to close with the timber recession in the early 1980s. Many mill and woods workers have transferred to new employment since then, and unemployment has dropped from 12 to 18 percent in the early 1980s to 6 to 9 percent by 1997.

Despite the relative economic health, local businesses and societal welfare are at risk if further mill closures or fishing restrictions occur. Declines in commercial fisheries have increased considerably in recent years. Due to this decline and the seasonal nature of commercial fishing, fisheries workers have been mildly receptive to employment promotion programmes. Timber workers, and many fisheries workers, exhibit peer pressure not to cooperate with new employment schemes. However, several local examples exist of timber/fisheries workers who transferred successfully to restaurant ownership or other service-sector entrepreneurship.

Expenditures by visitors to Curry County in 1991 amounted to over US$50 million, supporting approximately 1,000 jobs (Dean Runyan Associates 1994). Area tourism industry wages (full-time including tips) averaged US$16,000 per year, compared with US$25,000 per year in the lumber and wood products industry (Dean Runyan Associates 1994b). Despite the lower wage, the industry provides a high proportion of opportunities for entry-level workers, entrepreneurs, women and minorities (*ibid.*).

The most obvious natural feature of Curry County is its strikingly rugged and scenic coastline. The county is approximately 129 km long and 64 km wide. A high proportion of the Oregon coastline was protected as state park in the early part of the century. There is also little development between the three coastal communities, so most of the coastline is rural and scenic. There is less development inland due to private timber and public forest land. Five federally designated wild and scenic rivers (Elk, Rogue, Illinois, Chetco and North Fork Smith) and three designated wilderness areas (Grassy Knob, Wild Rogue and Kalmiopsis) lie within the county.

The Klamath Mountain Province of southwest Oregon and northwest California, encompassing the Siskiyou Mountains at its centre, is defined by its exceptional biotic and geological variety. Many locals do not realise that it contains the most species of coniferous tree in the world, as well as some of the largest outcrops of ultramafic, or serpentinic, bedrock. The hills east of Brookings also harbour the northern extent of the range of coast redwood.

Project origins

National forests have traditionally contributed to rural community economies through the forest products and visitors they generate. The USDA Forest Service, through the 1990 Farm Bill, received the authority and direction to begin working more directly with rural communities on economic diversification.

The Chetco Ranger District of the Siskiyou National Forest, with its office located in Brookings, began working on tourism development in partnership with the local chamber of commerce. Meeting facilitation skills were utilised to develop a local tourism development plan (Forbes 1993). One of the goals developed by the chamber's tourism committee was to promote tourism that capitalised on the area's outstanding natural attractions.

The facilitator of the plan, a Chetco Ranger District forester, attended a state-wide meeting on community and tourism development, in association with his new role in community assistance. Bob Harvey and Diane Kelsay of Egret Communications gave a presentation on ecotourism at the state-sponsored

meeting. They were subsequently invited to Curry County to assess its resources on the worldwide nature-based tourism market.

After meeting local agency heads and touring the surrounding countryside,
Egret Communications felt strongly that the county had a lot more to offer
than most counties as a nature-based tourism attraction. The relatively undeveloped character, botanical diversity, scenic coastline and the initial idea of
a temperate forest canopy walkway were thought to be marketable. Key local
leaders, including business leaders, state park managers, county commissioners
and mayors, were brought together to discuss the desire to initiate a project.
The timing of the proposal was good, coming several years after the contentious debate over the protection of the endangered spotted owl had peaked
and after a new forest plan (USDA/USDI 1994) was in place.

A key supporter was Republican county commissioner Rocky McVay, long-
time local resident and pro-timber lobbyist, who saw this as an opportunity
to generate revenue from the federal lands that were no longer available
for timber production. Michael Frazier, Chetco District Ranger (Siskiyou
National Forest), was also a key supporter who convened local leaders to assess
this opportunity. After initial questions were answered on possible impacts
on traditional industries, such as timbering and real estate development, consensus to proceed was reached.

An application was made to the Tri-County Regional Strategies Board,
which oversees economic development grant funds available from the sale
of state lottery tickets in Oregon. That programme funded the first phase of
this project for US$41,500. This allowed Egret Communications to meet a
wide variety of local business and social sectors, trading information on the
concepts of sustainable tourism development for information on local conditions and values.

The Northwest Economic Adjustment Initiative, part of President Clin-
ton's Northwest Forest Plan (USDA/USDI 1994), funded the second phase,
dealing with sustainability planning and product development (US$225,000)
(Dyer 1995). The initiative brought the Forest Service and ten other federal
agencies together with state and local governments to identify and fund priority economic diversity projects. The Oregon Economic Development Department funded the final phase, dealing with business training and marketing
(US$115,000).[1]

Sustainability planning

Egret Communications chose to use the term 'sustainable nature-based tourism' in defining the project. It felt that the term 'ecotourism' was perhaps
overused and had inappropriately spread to business ventures following the
values of traditional mass nature tourism, rather than a tourism highly sensitive to local cultures and environments. Sustainability was a better term to be
promoted as an alternative to traditional tourism development. Although
another buzzword, the term 'sustainable' was suitable for defining a key desired
value in the Curry County project.

Egret Communications had witnessed, through experience and contacts with foreign governments, the overdevelopment of destinations with the eventual decline of conditions that first attracted travellers. Just a few examples were Acapulco and Cancun in Mexico, Miami in the United States, and several islands in the Caribbean.

To try to prevent this occurring in Curry County, sustainability was built directly into the project through three mechanisms: (1) designing a self-sustaining, local, revenue-return system; (2) assessing local social values, through interviews and surveys, to avoid unacceptable impacts; and (3) assessing indicators of environmental and social change, through the limits of acceptable change (LAC) process (see Chapter 7), which was also used to predict and avoid unacceptable impacts.

Unacceptable impacts were defined as those impacts that would threaten the sustainability of the nature-based tourism venture: those leading to locals' resentment of visitors and/or deterioration of attractive resources. Altruistic and legal values also promoted concern for local species, social conditions and future generations. The term 'nature-based' tourism implies not only use of nature as an attraction, but a reciprocal relationship with nature.

Perhaps the unique value associated with the project can be seen in tourism products utilising visitors to perform restoration of fisheries and other habitats. This moved from merely avoiding unacceptable impacts to providing an increased benefit to the natural community, creating a 'reciprocal' relationship in development. The increased educational benefits from interpretation could also be classified as a benefit to the natural community.

Sustainability was also promoted indirectly through diversification of the existing county economy and its tourism sector. The underlying concept was that an economy more resilient to change would be more sustainable. The current problem of overdependence on one or two main cyclical industries (in this case timber and fishing), and on relatively unpredictable, drive-through tourism, would hopefully be mitigated through this project.

Key differences exist between this project and other commonly used models of development: (1) through the initial trading of information, it attempted to work through all sectors of the community, not just the tourism sector; (2) it followed steps to leave the county not only with tourism products, but with the mechanisms to sustain them without outside help (Harvey and Kelsay 1994).

Limits of acceptable change

A leading model for anticipating and preventing unacceptable environmental impacts is the LAC model (McCool and Stankey 1992). More commonly used in wilderness area planning, the LAC model was chosen for this project for its ability to anticipate and head off undesirable impacts. It recognises that impacts will occur from tourism development, and sets up a process to monitor changes before they reach an undesirable limit.

Dr Stephen McCool of the University of Montana has worked on recreational carrying capacity issues for over 25 years (McCool and Stankey 1992).

Under his leadership in a two-day workshop, the principles of LAC were taught to local national forest, state park, county and community officials so that each was familiar with the decision-making process and the rationale behind it. A county LAC team (made up of local people, and forest, park and other resource managers) and process was also established. Elizabeth Boo, ecotourism author and consultant formerly with the World Wildlife Fund, and Dr Steve Martin, professor in natural resource planning at Humboldt State University, assisted the team for several months after the initial training.

Indicative of the limited free time of local leaders, regular attendance dwindled from an initial group of thirty to approximately five to ten. The eventual working team was made up of a county planner, a county office manager, a port representative, a state parks representative, two national forest resource assistants, a city planning commission chair and two citizen planning team representatives. The team followed the nine-step LAC planning process (Cook 1996a):

1 *identify concerns and issues* – these included spreading benefits throughout the county, avoiding crowds in locals' favourite spots and identifying resources to be left alone and those to be mitigated.
2 *define and describe opportunity classes* – these resembled the Forest Service spectrum of recreation opportunity classes (primitive, semi-primitive motorised and non-motorised, semi-natural, rural and urban) (USDA Forest Service 1986). Descriptions of each class were based on amount of roads, trails, tourist experience, population density, lodging and eating facilities, land ownership and uses, landscape modification, and density of visitors. A county map of these zones has been developed for review and monitoring.
3 *select indicators of resource and social conditions* – indicators included number of new sustainable nature-based tourism (SNBT) businesses and jobs, occupancy rates at lodging facilities, average salary of SNBT jobs, recreational tourism revenue at resource management agencies, cost of living in Curry County, number of package tours by season, and growth rate of SNBT businesses.
4 *inventory resource and social conditions* – this was done by interviews and surveys conducted by Elizabeth Boo, described below.
5 *specify standards for resource and social indicators* – a matrix was developed showing direct and indirect indicators of change under the four categories of economic, environmental, social/quality of life and visitor experience (Cook 1996b).

Additionally, criteria that defined a sustainable nature-based tourism business were developed, with input solicited from local citizen groups. These criteria are not mandated for a business to participate in the project, but are ideal standards for monitoring. The key elements of an SNBT business included:

• utilises natural environments and provides opportunity for meaningful contact between people and the natural environment, either directly or indirectly;
• provides opportunity for meaningful interaction between outside tourists and local people;
• creates low or minimal impact on facilities, natural resources and the local social structure whereby these elements can be maintained indefinitely;
• employs at least 50 percent of its staff from Curry County, and if the business is individually owned staff would be located in the county;

• promotes and supports local goods and services; and
• visibly returns time, materials and/or money to local projects that maintain, restore or enhance the natural environment.

6 *identify alternative opportunity class allocations* – alternatives were not seen as necessary after step 2.

7 *identify management actions for each alternative* – actions were underway through tourism products being developed through local entrepreneurs. A non-profit tourism board was also formed to sustain the concepts of the project.

8 *evaluate and select an alternative* – no alternatives were seen as needed.

9 *implement actions and monitor conditions* – a monitoring plan still needs to be developed based on the matrix of indicators for each of four categories under step 5. Responsibility for monitoring falls to the various resource agencies and the non-profit tourism board.

Social assessment – surveys and interviews

Elizabeth Boo conducted this assessment (Boo 1995) using 600 random surveys and 71 selected interviews. A 32 percent return rate (180) was received on the surveys, which were mailed to registered voters in proportion to the number of people in each town. The results of the survey and interviews were interesting.

Only 20 percent of respondents were born in Oregon and only 3 percent were born in Curry County. Retirees comprised 42 percent of respondents, and only 20 percent of these were born in Oregon. Fishing was felt to be by far the most popular attraction for visitors, and 25 percent had jobs providing products and services to tourists.

Overcrowding was not yet perceived to be a big problem and few perceived negative impacts were expected from increased tourism (with the exception of increased traffic and prices). Most respondents felt that the small town atmosphere of Curry County was important to maintain.

The creation of new tourism businesses and jobs was seen to be the main benefit that expanded tourism would bring. The new businesses could also be enjoyed by locals, and could generate community pride in showing outsiders locally unique features. However, the majority felt that the new jobs would not be at a family-wage level.

A commonly cited problem was resident attitudes towards visitors, which were often discourteous and indifferent. Fisheries and timber workers would have trouble changing to lower-paying service jobs, which they saw as 'babysitting'. Some respondents expressed fear of bringing more 'tree-huggers' to the area who might further threaten traditional livelihoods. Respondents suggested that training was needed in tourism-related service, and that more year-round activities were needed. There was strong concern to involve locals in every step of the process and keeping the lines of communication open.

Revenue-return system

Dr Peter Forsberg (World Travel and Tourism Tax Policy Center, Michigan State University) worked with Egret Communications on-site to set up policies for recovering revenues from visitors and channelling them to accomplish business and resource management needs (Forsberg, Harvey and Kelsay 1996). In the past, most Curry County tourism expenditures went on lodgings. Private camping grounds and hotels/motels predominated in generating jobs and payroll. However, only a portion of the room taxes were used for tourism promotion. Gold Beach used 79 percent of its room assessment revenues for tourism advertising. Brookings used only 25 percent for advertising, while Port Orford allocated all of its room taxes to its general fund. By contrast, neighbouring Coos County to the north allocated 100 percent of its room taxes to marketing.

It has been proposed that Curry County levy a 6 percent county-wide room tax. This would generate six times the current room tax revenues, bringing in US$3 million a year. Funds would be used for resource management, new product development, product improvement and marketing. The 6 percent rate is thought to strike the best balance of demand and revenue.

The tax is proposed to be applied to all transient lodging facilities – hotels, motels, bed and breakfasts, RV (recreational vehicle) parks, private, state and federal camping grounds, and current state and federal recreation rental facilities such as fire lookouts, cabins and yurts. Public and private camping grounds capture a major part of the revenue and are important to include in the county room tax proposal. Other recreation users that may contribute to the tourism enhancement fund, called the Heritage Assessment fund, are those associated with the product development section of this project: mountain bike tours, intertidal explorations, salmon stream enhancement, crabbing tours, sport diving, photographer and artist services, and the proposed elevated forest walkway.

An interesting revenue-return proposal is to place a flat tax on new residential development. Since substantial new development may threaten the tourism attraction of a relatively undeveloped area, a special tax on it would help to fund sustainable development initiatives and mitigations.

Forcing recreational vehicle owners (trailer campers or RVs) to use proper camping grounds was also recommended to generate local revenue in proper exchange for their use of local services. Signing at alternative route intersections could discourage use of the county simply as a means of reaching other destinations.

A non-profit corporation was recommended to oversee dispersal of the Heritage Assessment funds, as it would be more flexible and less burdened by government regulations. Most funds in the early years would go to marketing until a visitor base is built – then be transferred to resource enhancement.

Product development

The development of package tourism products, not previously offered in the local area, was seen as a key to generating revenue and stimulating new businesses. Local entrepreneurship, with assistance from Egret Communications and visiting experts, developed the products described in Table 10.1. Marketing brochures have integrated these new products into regional advertising

Table 10.1 New tourism products in Curry County

Mountain bike tours	this product focused on developing quality mountain bike tour routes of easy to moderate difficulty. Using existing roads was emphasised over creating new trails. One new business has purchased bikes and a support van, and will start with half-day tours through unique national forest habitats close to the Brookings–Harbor area.
River kayak tours	these more adventurous tours will be provided through the 'wild' section of the Chetco Wild and Scenic River. Access will be by horsepacking in from a mountaintop wilderness trailhead. This tour will feature outstanding scenery, floating through 35-foot deep pools of emerald-green water and under 50-foot rock ledges.
Lighthouse tours	a new operator is leading historical tours of Cape Blanco lighthouse, the oldest continually operating lighthouse on the Oregon coast. This north county site is also an excellent location for bird-watching and storm-watching. During winter storms, it is often the site of the highest winds in Oregon, ranging from 40 to 100 km per hour.
Cranberry bog tours	promotion of locally grown products has been recommended as a means of capturing visitor interest and revenues. Commercial cranberry bogs at the north end of the county have instituted new tours, and though most berries produced locally are for juices, marketing of additional cranberry products (preserves, baked goods) to visitors will occur.
Stream restoration work projects	this product will involve visitors in rehabilitating streams and improving salmon habitat. Most work is expected to include planting conifers and cutting understory brush and overstory alders that had colonised the forest after logging and flood disturbances. The conifers provide better long-term habitat through nutrient diversity and large logs that fall in the creeks to create pools and regulate sediment. Salmon spawning surveys, an especially rewarding experience, may also be included.
Offshore/marine tours	one commercial fishing operator has proposed tourism activities out of Port Orford. Opportunities may include bird-watching trips on fishing boats, catch-and-release fishing trips, day trips with commercial fishers, recreational crabbing, whale watching, and diving.
Photographer/artist workshops	this product focused not only on utilising Curry County's scenery as subject material, but through one new business, also developing visiting artists' deeper sense of history and place. Information was compiled on scenic sites, networking with photography/artist groups and packaging tours. Marketing was developed through creation of a booklet on local photography and art opportunities.
Elevated forest walkway	this involved site selection and design of a walkway suspended in the canopy of an old-growth forest stand. The product would capitalise on a combination of recent research interest in temperate forest canopies and visitor interest in old-growth forests. A partnership of private and public funding has been sought for the proposed $19 million project, which includes operating funds for the first two years. Local and outside interest has been considerable. Some funds were used to bring in architects experienced in designing such walkways that currently attract visitors to tropical forest sites.

efforts. These activities are likely to be mixed in as one or two-day events within a week-long package (Harvey and Kelsay 1996a).

An interpretive centre would be located at a centralised, scenic oceanside site about 24 km north of the walkway sites, to distribute visitors throughout the county better. The interpretive centre would have interactive exhibits. One possible exhibit would link with a non-obtrusive camera located near an osprey nest on the Rogue River. Visitors could manipulate a 'joystick' to move the camera and view both the osprey and the wild and scenic river, with its passing river otters, waterfowl, deer and black bears. Partnerships to set up links in big-city museums such as the American Museum of Natural History (New York) have been discussed. This would generate further interest in visiting the county.

Business training

This aspect of the project involved working closely with Southwest Oregon Community College (SWOCC) to provide business training and assistance for entrepreneurs, existing businesses and guides (Forbes 1996). A diverse series of workshops took place in a compacted time frame in the autumn of 1996. Workshop subjects included:

- sustainable tourism opportunities for new entrepreneurs and existing businesses;
- international marketing/working within the travel industry;
- tourism dynamics – balancing cooperation and competition;
- travel agent familiarisation tours;
- insurance;
- techniques for guiding nature field experiences;
- guide partnerships – businesses and community;
- the art of interpretation – mixing fun, reverence and learning;
- preparing to host nature travellers;
- nature lodge operation and management; and
- training and assistance for start-up businesses.

Continual training for local businesses, utilising local natural resource and business specialists, is being set up through SWOCC. One of the most difficult aspects of starting up new businesses has been the pricing of tours, as relationships with distant tour operators and agents could be damaged by dropping or raising prices afterwards. Most tours started with a price around US$70/half-day per person, including lunch.

Marketing Curry County

The target market for Curry County is nature travellers. Environment-based tourism is predicted to grow at 25 to 30 percent, compared with a 8 percent growth rate for world tourism (Harvey and Kelsay 1996b). Most accounts report this as a trend rather than a fad. Increasingly urban environments create demand for nature travel. This market can be reached with careful niche advertising in publications or television programmes linked to environment, natural history or science. It also can be reached by creating small package tours marketed through nature travel providers. The county is marketable to inland, urban residents of the US Midwest and East, who find the coastline and rugged scenery a unique contrast to their surroundings. Previous marketing has focused on visitors within a day's drive, but future marketing will focus on air travellers and those attracted to package tours.

A marketing strategy was developed for the county (Harvey and Kelsay 1996a). This strategy addressed eight key points:

1 establishing a group of focused patrons (a niche even smaller than targeted patrons) to make best use of advertising resources;
2 establishing an image through brochures and photos that highlight unique attractions – scenic, undeveloped coastlines and mountains and botanical wonders;
3 using marketing to link existing and new tourism products;
4 using marketing to expand the tourism season;
5 developing marketing that appropriately matches the type of visitor to the product and orients them to the area;
6 developing marketing that meets the unique goals of the new products associated with this project;
7 developing marketing that reflects both the short-term goals (starting the flow of visitors) and long-term goals (building partnerships to attract repeat visitors and link products); and
8 using several diverse businesses to 'crack' the market was thought to be more effective than using a single entity.

The marketing effort has already leveraged limited advertising funds by networking with travel writers and editors, travel agents, and other representatives of groups likely to send visitors to Curry County. Plans for a large marketing event were changed to accommodate the current emphasis on one-on-one familiarisation tours for travel writers and operators. Press releases and articles were distributed on the proposed elevated forest walkway, and on the project in general. Attractive brochures were developed, using the theme 'Oregon's Siskiyou Coast'. This theme reflects the rugged coastal nature of Oregon combined with the unique botanical and geological features of the surrounding Siskiyou Mountains.

Conclusions

The Northwest Economic Adjustment Initiative (NWEAI), with its funding associated with the US President's 1994 Northwest Forest Plan (USDA/USDI 1994), provided the opportunity to experiment with the leading concepts of sustainable tourism in a rural North American setting. Such a large effort would not be appropriate in many of the counties affected by the decline of federal timber revenues. Curry County has the unique natural attractions of a Pacific Ocean coastline and botanical diversity to market.

Much of the credit for this effort goes to the elected county commissioners and their staff, who had the foresight to back such a unique project. Proponents within the rural development community see this as a different type of initiative from other NWEAI projects. Many NWEAI projects over the first years of their life (1992–96) have been 'one-shot' infrastructure improvements for mills or municipalities. Although beneficial, these have not sought to develop and sustain a new sector of the economy as this project has.

The project is nearing the end of the four-part initiative developed by Egret Communications as tourism advisors. Although Egret has moved to Curry County and will continue to be a resource, local agencies and citizens were to take over the project in the autumn of 1997. A citizens' LAC committee has been established, as well as a solid sector of local support.

Initial conflict occurred between the consultants' desire to slowly develop high-quality, marketable products and educate locals and the desire of existing groups (chamber of commerce, traditional tourism operators, local environmental groups) for hands-on ownership of the project with more immediate results. Some locals have grown impatient with the expense and time involved in setting up the mechanisms for sustainable tourism. Scepticism has surfaced as to the ability to attract visitors, and the size of the grant that paid for the planning process.

Future projects of this nature might focus on the patience needed to generate sustainable tourism businesses. Earlier ownership of the process by traditional local groups could be encouraged. The project advisors also discovered more bureaucratic hurdles, such as lengthy permitting processes and complex grant restrictions, than are found in less developed countries such as Belize.

Despite the length of time involved in planning (three years, 1994–97), nine new business ventures have started, with 30 more in the idea/start-up stage, and a climate of cooperative marketing has been established between them and several existing businesses. Local officials have been very intrigued by the possibility of the US$19 million elevated forest walkway and interpretive centre. A non-profit county tourism board has been set up, as well as an elevated forest walkway oversight board.

Two keys to future success are, first, the cooperative survival and growth of new businesses and products; and, second, realisation by a broad spectrum of county leaders that conservation and business can help each other to thrive. An imperfect blend of traditional and sustainable development is likely to prevail in the county in the near future. With faithful attention to the concepts and mechanisms of the project, the theme of 'enlightened self-interest' may grow, blending reciprocal and sustainable development.

The advisors, Egret Communications, deserve credit for persevering through controversy to develop marketing publications, coach new tourism businesses, and bring in world leaders in sustainable tourism as advisors. They have introduced a way to mix nature conservation and business that has not been seen before on Oregon's south coast.

Endnote

1 The project described in this chapter was funded with grants from the Coos Curry Douglas Regional Strategies Program, the US Department of

Agriculture (USDA) Forest Service Rural Economic Diversification Program (administered by the State of Oregon Economic Development Department), and the USDA Forest Service Rural Community Assistance Program. Correspondence may be sent to Chetco Ranger District, Siskiyou National Forest, 555 Fifth St, Brookings, OR 97415. Additional contacts: Egret Communications, PO Box Q, Port Orford, OR 97465; Southwest Oregon Community College (SWOCC), 420 Alder, Brookings, OR 97415, USA.

Participants in the workshops and other phases of the Curry County project include Bob Harvey and Diane Kelsay, Egret Communications; Tom Grasse, International Expeditions; Sally Sederstrom, Oregon Tourism Commission; Jim Bouley, Curry County Business Development Center, SWOCC; Chuck Box, Rocky Mountain International; Gene Bryan, Wyoming Department of Commerce and board member, US Travel Industry Association; Chuck Coon, Wyoming Division of Tourism; Brian Corcoran, Cramer and Giles Insurance; Bart Mickler, Maya Mountain Lodge and Educational Field Station, Belize; Rex Poulsen, Great Outdoor Shop, Pinedale, Wyoming; Charlie Love, Western Wyoming Community College; David Andersen, The Andersen Group Architects Ltd; Elizabeth Boo, ecotourism author and consultant; Steve Martin, Humboldt State University; Stephen McCool, University of Montana; Peter Forsberg, World Travel and Tourism Tax Policy Center, Michigan State University.

Chapter 11

Public transport and sustainable tourism: the case of the Devon and Cornwall Rail Partnership

Clive Charlton

This chapter examines the Devon and Cornwall Rail Partnership, a collaborative venture in southwest England that has sought to foster sustainable rural recreation and tourism while maintaining the viability of local public transport. It raises issues relating to the relationship between rural recreation and transport and also explores the implications of the partnership as a vehicle for policy implementation in sustainable tourism and transport.

Although attachment to the private car remains formidable (Barrett 1995), there is now a widespread consensus that unfettered growth in road transport is undesirable and unsustainable. The Royal Commission on Environment and Pollution (Department of the Environment (DoE) 1994) highlighted the environmental consequences of road traffic growth, especially in terms of air pollution, and set out a broad vision of the changes that would be needed for a more sustainable transport policy (Banister 1995).

In parallel with its prominence in the transport debate, sustainability has been inserted into the mainstream agenda for tourism policy and planning (Mowforth and Munt 1997). The influential report *Maintaining the Balance* (ETB 1991) stressed that sustainable tourism should seek to balance the interests of the tourist, the local community and the environment. Similarly, the Countryside Commission (1995) suggests that tourism should aim to 'minimise impact on the global environment', 'sustain the local environment' and 'sustain the host community and the visitor'. Sustainability in rural tourism is seen as 'an approach, not a finite set of activities' that should be pursued at all levels – national, regional and local (South West Coast Path Steering Group 1996).

The transport implications of rural tourism and recreation

The past two decades have seen rising awareness of the environmental consequences of the private car for recreation and tourism in Britain (Cullinane *et al.* 1996). The Countryside Commission (1992) warned of the dangers implicit in the forecast that road traffic mileage would grow nationally by between 83 and 142 percent from 1988 to 2025. Although levels of car

132

ownership in rural areas are generally high, much of the burden on rural roads is generated by urban-based motorists; at least 55 percent of recent rural road traffic growth has been leisure-related (Barrett 1995). The pressures are accentuated by the very uneven distribution of recreational traffic in time and space, a problem particularly apparent in the national parks, in which recent surveys revealed that 91 percent of visitors arrived by car, and 'traffic' was the most widely perceived problem (Countryside Commission 1996b). Among the particular problems of tourist traffic are 'traffic congestion, overcrowding due to traffic levels, inappropriate development at tourist sites and conflicts with local communities' as well as 'poor parking provision, disruption to local road users, and visual intrusion' (Countryside Commission 1992: 7).

A number of strands of policy and practice have woven a diverse web of what might be termed the 'new rural tourism', in which the range of tourist destinations, recreational activities and associated journey patterns has become more varied and complex, bringing the impacts of additional road traffic to more rural locations and environments (Clark *et al.* 1994). Tourism and recreation are valued as a source of new jobs and income in the face of long-term structural changes in rural economies, as typified by the support for farm tourism (Rural Development Commission (RDC) 1992; MAFF 1994). Tourism is also seen as a way of sustaining rural services, such as village shops and country pubs, thereby providing further reinforcement for rural communities. At the same time, the market for recreation has become more sophisticated, with a growing preference for specialised interests and activities, many of which can be supplied in rural settings (Countryside Commission 1996). A further dimension is a general desire by many for a closer engagement with 'nature' and the 'authentic', which, it is perceived, can be discovered in the wider countryside. Hence the growth in visits to nature trails, rural museums, heritage centres and restored buildings, as well as in active pursuits such as walking, climbing, golf and recreational cycling. There is also more tourist accommodation in remote rural locations, ranging from individual conversions of cottages and former agricultural buildings to substantial holiday villages and complexes, such as those operated by Center Parcs (Clark *et al.* 1994).

Those who claim that the new rural tourism has 'green' credentials point to such characteristics as a higher level of environmental consciousness and 'awareness' on the part of both consumer and producer of rural tourism, the dispersal of recreational activity in time and space, and the sustenance of local communities both economically and through the 'celebration' of cultural traditions. However, considerably more traffic overall enters rural road networks, on a greater range of links and at many more nodes than in the past. Most destinations are isolated from public transport routes, and potential demand for the latter is spatially dispersed as well as more varied in time, given the complexity of preferences for where and when people seek their rural experiences. Many recreational activities demand the transportation of bulky items, ranging from the elaborate ingredients of the modern picnic and barbecue to sports equipment such as golf clubs, surfboards, dinghies and hang gliders. The popularity of rural cycle trails depends fairly heavily on the carriage by car of well-loaded bike racks from urban areas. Traffic patterns are also complicated by the many irregular flows generated by countryside events, such as fairs and festivals, many of which have been stimulated and marketed as part of sustainable tourism initiatives.

Transport and rural recreation – an overview of policies

Policy makers have responded to the pressures from recreational traffic in a variety of ways. Broadly, the tactics available are the control and management of private road vehicles, and reinforcement of the role of public transport. National parks have been important pioneers of rural traffic management policies in Britain. Examples include the Lake District Traffic Management Project (DoE 1995), the Dartmoor National Park Traffic Management Strategy (Dartmoor National Park Authority 1994; Fewings 1996), and the Upper Derwent Valley project in the Peak District (Lumsden 1992; House of Commons Environment Committee 1995). Traffic problems in rural settlements are normally handled through traffic diversion, parking restrictions and parking provision outside the settlements concerned (as at Polperro in Cornwall and Avebury in Wiltshire).

However, more stringent restrictions on vehicle movements can meet hostility from local residents and businesses. When the Dartmoor National Park Authority attempted to introduce a traffic management scheme in the vicinity of Burrator Reservoir on the southwestern fringes of Dartmoor, which is particularly popular with day visitors from nearby Plymouth (Cullinane *et al.* 1996), there were vigorous local protests, to the extent that the original proposals were withdrawn. Such interventions demand careful local consultation and a thorough understanding of the local situation (Countryside Commission 1995). Policies for further traffic restraint remain on the agenda, however; the Countryside Commission is taking forward its concept of 'quiet roads' in rural areas, on which priority will be given to walkers, cyclists and horse riders (DoE 1994; Countryside Commission 1996).

It is widely accepted that rural traffic restraint should be complemented by adequate alternative public transport, and there have been many attempts to offer bus or train services to the rural recreation market. However, decisions to develop, market and support such recreational public transport operations are made not only for reasons of environmental sustainability (Countryside Commission 1987). Social and economic considerations in both urban and rural areas have been used to justify support for leisure-focused services. A key goal has been to provide countryside access for urban residents. In 1992–94, 34 percent of all urban households in the UK were without a car, compared with only 22 percent in rural areas (Social Trends 1996). The elderly, the young, women, those on low incomes and single-parent households are generally the least mobile: for many, 'rural tourism' is not a ready option. In addition, the range of regularly scheduled services in rural areas tends to be very sparse and poorly publicised, with timetables geared to serve the needs of travellers going to cities, rather than those seeking to escape them.

Although levels of car ownership are generally higher in rural areas, a significant minority of rural dwellers suffer severe accessibility problems (Cloke *et al.* 1995). This 'travel poverty' (Root *et al.* 1996) has been exacerbated by the long-established cycle of rising car use by the mobile majority, and decline in public transport patronage and services for the minority who still depend on them (DoE 1995). If recreational visitors can be persuaded to travel on existing bus or train services, existing rural transport systems may be more sustainable. Where new services are operated primarily to provide access to the countryside, improved mobility may be a spin-off available to rural inhabitants.

Many public transport operations in rural areas act not only as a means to recreational activity but also as a leisure experience in their own right (Countryside Commission 1995), for example the many preserved railway operations in Britain, such as the narrow-gauge 'Great Little Trains of Wales' and the North Yorks Moors line.

Any review of recent experience with traffic management and rural recreational public transport projects in Britain must acknowledge the rather harsh reality that they have yet to make a major impact on the growth of private car traffic. Many schemes have been unsuccessful, and have been abandoned or modified after a few years (Cullinane *et al.* 1996). Eaton and Holding (1996: 64) conclude that, in national parks, 'little has been achieved in influencing the choice of [transportation] mode' through the introduction of new public transport schemes. They are critical of the lack of clear objectives for such initiatives and, especially, of adequate mechanisms for monitoring and measuring effectiveness.

Partnerships in rural tourism and development

There have been significant changes in the nature of regional and local management in Britain, characterised by a relative shift from direct involvement by local and central government to a more complex, shifting organisational landscape. More power, resources and responsibility have passed to the private and voluntary sectors, as well as to the numerous quasi-autonomous government agencies ('quangos'), such as urban development corporations and training and education councils (TECs). Funding has become more competitive, with 'packages' of initiatives bidding for resources through such frameworks as City and Regional Challenge, as well as regional development funds available from the European Union.

Prominent within what might be characterised as a 'post-Fordist' mode of local governance (Goodwin and Painter 1996) have been many partnerships and collaborative projects, with specific objectives and areas of operation, and usually limited duration (Prior 1996). Justifications for the partnership approach have included a logical response to budgetary constraints through the pooling of funding and other resources; greater cohesion and avoidance of duplication in the design and conduct of policy; advantages in combined action when seeking external funding; the synergy and mutual support derived from pooled expertise and enthusiasm; a greater range of 'contacts'; greater freedom and stimulus for innovation outside the structures of 'mainstream' agencies; and the ability to foster closer linkages with local communities (Curry and Pack 1994).

Although especially evident in urban development and regeneration (Bailey 1996; Lowndes *et al.* 1997), partnership$ are also widespread in the rural context, often in connection with rural tourism and countryside management (Countryside Commission 1995). Several partnerships have been formed with a combined transport and recreation focus; besides the Devon and Cornwall Rail Partnership case study that follows, examples include initiatives centred on the Settle and Carlisle railway, the Esk Valley railway in North Yorkshire (Buxton 1997), the Hope Valley line in the Peak District, and the Heart of Wales line between Swansea and Shrewsbury (Wilson 1997).

The Devon and Cornwall Rail Partnership

Devon and Cornwall are the two most visited counties in the West Country, the most significant region in terms of tourist spending outside London (RDC 1997). Tourist road traffic is especially severe in summer, although seasonality has been declining, so that traffic problems occur over a longer period and in more places than in the past. Recreational activities in the region generate a complex pattern of mobility – trips to beaches and tourist attractions, exploration of the deeper countryside, and visits to villages and towns. Besides the main distributor roads, the pressures fall on local routes, many of which are picturesque and ancient, but also narrow and relatively hazardous, even at low speeds. The impacts include slower journeys, streets crowded with vehicles, overstretched parking facilities, awkward and illegal roadside parking, noise, and air pollution. The effects are felt in popular resorts such as St Ives and Newquay, as well as in the more remote, smaller-scale locations that underpin much of the area's appeal for the visitor. Local residents tend to view themselves as victims (rather than contributors to the process), which fosters hostility towards 'tourism', especially in the summer season. Congestion also places an additional cost burden on local businesses (Council for the Protection of Rural England 1993) and becomes an abiding memory of the holiday experience that discourages repeat visits. The sustainability of the region's tourist industry may then be questioned from an economic viewpoint, as well as in the conventional environmental sense.

As elsewhere, there is also an awareness of the responsibility to meet the needs of residents without access to a car, and of those who prefer not to use their cars for leisure trips (Cornwall County Council 1996). A significant number of tourists, albeit a minority, come to the region by public transport, by choice or necessity. They also depend on the local public transport system for their mobility. However, the low demand across much of the region means that many rural areas have, at best, a rudimentary level of service, certainly insufficient to act as an attractive alternative to the car for leisure trips.

The Devon and Cornwall Rail Partnership is an attempt to achieve sustainability in transport provision, and also rural recreation and tourism. It also illustrates some of the implications of the partnership mode of organisation. The partnership began operating in 1991, with the fundamental aim of encouraging greater use of the railway branch lines in Devon and Cornwall for leisure and recreation (Figure 11.1, overleaf). Each of the lines is distinctive in terms of markets served and operational characteristics. For example, the Tamar Valley line provides access to Plymouth for communities with very poor alternative road links; the Looe Valley line provides a scenic, but decidedly rustic, connection for holidaymakers and local residents between the popular small resort of Looe and the market town of Liskeard, which has a main line rail connection; and the line between Par and Newquay has a very seasonal market of tourist traffic to and from Cornwall's major seaside resort. All the lines have some relevance, actual and potential, to the tourist industry of the two counties, and all, to varying degrees, provide a necessary public transport service for local communities.

The partnership arose through a coincidence of policy concerns among the sponsoring organisations. It could be argued that a central aspect of its evolution was the inability, or reluctance, of the railway company to undertake the necessary investment in development and marketing. However, the partner organisations have additional priorities beyond those of the train operator, as well as a wider appreciation of the local and regional context of the railway lines, especially as regards tourism and recreation, and the needs and potential of the local community and environment. Besides the basic purpose of expanding the market for travel on the branch lines, the partnership has several other aims (Devon and Cornwall Rail Partnership 1996):

- to boost patronage of the existing train services, thereby improving their long-term viability and maintaining the mobility of residents on the lines;
- to reduce the impact of recreational vehicle traffic on the rural environment;
- to provide a means for tourists and local residents without access to cars to reach the countryside for recreation; and
- to bring economic benefits to communities served by the railway through the spending of visitors arriving by train.

The partners in the first three-year phase of the partnership, from 1991 to 1994, were:

- Cornwall County Council
- Devon County Council
- The Countryside Commission
- The Rural Development Commission
- The University of Plymouth
- Regional Railways, South Wales and West

Although all partners supported the project aims, they differed in terms of their priorities. For instance, the project met the Countryside Commission's desire to widen access to rural recreation opportunities and to reduce the damaging impacts of road traffic growth in rural areas, while the RDC recognised the potential for economic and social development in rural communities through additional tourist revenue from leisure rail travellers. Devon County

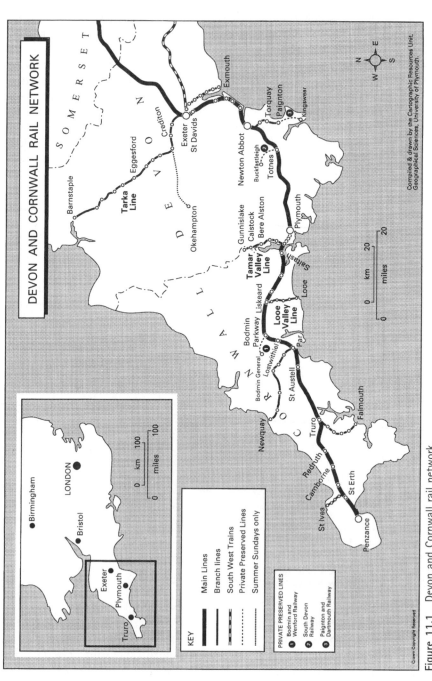

Figure 11.1 Devon and Cornwall rail network.

KEY

Main Lines

Branch lines

South West Trains

Private Preserved Lines

Summer Sundays only

PRIVATE PRESERVED LINES
1 Bodmin and Wenford Railway
2 South Devon Railway
3 Paignton and Dartmouth Railway

DEVON AND CORNWALL RAIL NETWORK

Compiled & drawn by the Cartographic Resources Unit, Geographical Sciences, University of Plymouth.

Crown Copyright Reserved

Council particularly valued the project's contribution to the protection of rural public transport services: the county's input to the partnership has come from its well-established and active Transport Co-ordination Centre. Cornwall County Council's support rested principally on two departments: the Highways Section, which had begun to accept that alternatives to road transport should be fostered in the face of traffic growth, and the Countryside Management Service, which was in a particularly dynamic and expanding phase during the earlier years of the project. The University of Plymouth, which provided the partnership with a base, saw the venture as complying with its mission to serve its local region, as well as an opportunity to integrate a 'live' project with research and teaching.

The initial three-year phase of the partnership was followed by a second, from 1994 to 1997. Although the Countryside Commission was unable to remain a core contributor to the project, in compliance with its general policy against repeat project funding, it was possible to retain the support of the RDC, and to add the Dartmoor National Park Authority and Plymouth City Council as partners.

The partnership's operations

The partnership is staffed by a full-time officer with a part-time assistant. Broad strategic control of the partnership is vested in a steering group, which meets twice a year, and includes local authority councillors among its membership. Advisory support is provided by an 'officer group' of representatives from the partners, who generally have a long-term commitment to the project. The core budget of £32,000 (approximately $US53,000) per annum for the first phase and £43,000 (approximately US$70,000) in the second is scarcely sufficient in itself to deliver any significant action. In practice, the partnership has operated as a catalyst, coordinator and supporter of collaborative initiatives that draw additional funding and support from a network of alliances within the region and beyond. Such organisational complexity has become increasingly standard in the management of rural tourism and rural development, reflecting the wider shifts in the nature of regional and local governance in the 1990s. In addition, the partnership has pursued its overall aims of increasing leisure-related use of the local rail network through a broad-based programme of initiatives, the most important of which are detailed in Table 11.1 (overleaf).

Initiatives on individual branch lines

Much of the partnership's work has been conducted through 'branch line specific' projects, managed by the Rail Partnership Officer but guided by line promotion groups and funded through additional, separate budgets. Each of these line promotion groups has drawn in organisations additional to the 'core' sponsors of the main partnership. For example, the Tamar Valley line group includes the two local district councils (Caradon and West Devon), the National Trust, and a representative of the small and medium-sized hotels in

Table 11.1 Partnership initiatives

- Publicity and interpretive materials: 'line guide' brochures for branch lines; publicity maps for the regional network; posters on stations with information on the communities served; simplified timetable and 'day out' sheets.
- Development and publicity of bus links serving stations, as at Gunnislake and Bere Alston in the Tamar Valley, and Eggesford on the Tarka Line.
- Recreational links from stations: cycle hire facilities and routes (at Barnstaple and Eggesford on the Tarka Line); walks from the railway, using self-guided leaflets and guided walk events.
- Station enhancements: repainting; improved signing, interpretive posters, and timetable displays; creation of murals in conjunction with local schools.
- Sunday rover ticket on the Tamar Valley line; special promotional fares.
- Publicity at local events, in the local and national media, in shops, town centres and libraries.
- Liaison with the tourist industry in the southwest: incorporation of information on rail travel in general promotional materials; negotiation of discounts at tourist attractions for those arriving by train; 'educational' trips for tourist information centre staff and journalists.
- Educational work: rail excursions for local schools and colleges; production of 'education packs' and materials.
- Research in support of the partnership's aims.

Plymouth (which appreciate the line's potential as an attraction for their guests). With each of the branch lines, there have been close links between the partnership and other rural recreation and tourism projects in the vicinity. The Tarka Line, from Exeter to Barnstaple, shares its name with the Tarka Project (Countryside Commission 1995), which has used the name of Henry Williamson's fictional otter to market 'green tourism' and undertake conservation initiatives in north and mid Devon. Part of the railway was incorporated into the route of the circular 'Tarka Trail' walking route. The Looe Valley line was promoted as part of 'Project Explore', another partnership established to develop and market sustainable rural tourism in the hinterland of Looe. Similarly, there has been close cooperation with the Tamar Valley Countryside Management Service, particularly in the provision of guided walks and other events located and timed to allow easy access by rail.

The Tamar Valley line has also become an important element of the 'Dartmoor Sunday Network' of public transport services organised and part-funded by Devon County Council and the Dartmoor National Park Authority. Since 1990, an extensive summer bus network has been developed linking points in the National Park to nearby urban centres on Sundays. The bus operations aim primarily to provide an alternative means of access to Dartmoor for tourists and residents of the region, but communities in the park itself now have a much enhanced level of mobility, even if not on a daily basis. A strong marketing feature of the Dartmoor Network has been the 'Sunday rover' ticket, which provides flexible, unlimited travel on all the bus services on summer Sundays, with particularly attractive discounts for 'family' groups.

'Ride and Ramble' events on Dartmoor and in the Tamar Valley have been offered for a number of seasons, and vintage buses chartered to operate two of the leisure routes.

As part of the Dartmoor Network, a Sunday train service from Plymouth to Gunnislake is chartered by the line's working group, on which the rover ticket is valid. Four different bus routes are timed to meet trains at Gunnislake, offering a very wide of leisure travel opportunities. The Rail Partnership is responsible for organising volunteer assistants who travel on these trains, supplying information on timetables and the availability of leisure facilities, and making suggestions for visits to attractions and walks accessible from the railway. Many who use the Sunday service are city dwellers who rarely use either the railway or the countryside.

Although much of the work of the partnership has been directed at individual branch lines, there have also been initiatives with a region-wide coverage. For several seasons, a relatively high-quality map-guide entitled 'Explore Devon and Cornwall' was produced, with the aid of funding from the two county councils, the Cornwall Tourist Board, all three train companies operating through Devon and Cornwall, and also three preserved steam railway organisations. The intention was to publicise the region's rail network in general, and the recreational opportunities that can be accessed from it, with distribution via tourist information outlets in Devon and Cornwall, and also beyond the immediate region. In 1997, a more ambitious publication was issued, featuring a map with comprehensive and detailed coverage of public transport available to leisure travellers in the two counties, a series of suggestions for 'car-free days out', and information on rover tickets. Targeted at the British domestic holiday market, this publicity material represented a particularly impressive degree of coordination by the partnership, especially in achieving the collaboration of the principal bus and train operators serving the region, as well as acquiring funding derived ultimately from the European Regional Development Fund.

The partnership has also produced a leaflet to promote the use of the branch lines to overseas visitors, written in six languages, including Japanese. This has been distributed in key gateway tourist information centres in Britain, as well as the countries whose languages are featured, through a partnership with British Rail International (Devon and Cornwall Rail Partnership 1996).

The achievements of the Devon and Cornwall Rail Partnership —

Comprehensive monitoring and evaluation of such a multifaceted enterprise as the Devon and Cornwall Rail Partnership would be a demanding task. The project was established with a number of related broad objectives against which performance could be judged. However, the partnership's activities cannot be isolated from the other projects and programmes with which it is enmeshed. They are subject to influences over which the partnership has very little control, such as the performance of the regional tourist industry and railway system as a whole, and the condition of the economy and consumer spending. Some key information, such as details of rail passenger traffic and revenue, has become especially elusive for commercial reasons under the new regime of rail privatisation. Nevertheless, it is possible to put forward a number of comments that confirm the partnership's status as a worthwhile experience in sustainable tourism management, and also some questions and qualifications that are necessary for a realistic evaluation.

There is some confidence that the core aim of attracting additional leisure-related patronage to the branch lines of Devon and Cornwall has been achieved – or more specifically that the background decline in the 'non-leisure' local rail market has been considerably offset by gains in recreational travel. In the first phase of the project, passenger numbers on the Tamar Valley line increased by around 7 percent per annum, although this growth could not be sustained in the second phase. The Looe Valley similarly saw a significant rise in patronage between 1992 and 1994. Research conducted by the partnership suggested that between £0.45 million and £0.82 million is spent annually by visitors travelling to Looe by train (Mowforth and Charlton 1996). A survey conducted on the Tamar Valley, Looe Valley and Tarka lines in 1992 (Charlton 1992) revealed that just over half of the travellers outside commuting periods could have used a car but had opted to use the train. Around 25 percent specified their concern to avoid traffic and parking problems as a reason why they were making the rail journey, and 45 percent felt that the train journey was 'more enjoyable'.

The performance of the partnership has earned it a strong reputation nationally and within the region. It has established extensive linkages with the public, private and voluntary sectors in Devon and Cornwall, and has thereby made a significant contribution to furthering the cause of sustainable rural tourism and to a more positive consciousness of the local rail network. The partnership enjoys the esteem of local authorities, an important factor in ensuring its continuation in an era when some comparable projects have proved short-lived. On a wider stage, the partnership has been recognised as a 'good practice' case study in several key national reports on rural issues (Countryside Commission 1995; DoE 1995). The Devon and Cornwall experience has also lent support to the adoption of the partnership model for railway lines elsewhere in Britain, such as the Heart of Wales and the Esk Valley lines.

A number of factors help to explain this relative success. The independence and flexibility of operating at arm's length from its sponsors, with a full-time partnership officer, seems important. This gives a neutrality and freedom from bureaucratic and political constraints that can inhibit mainstream public sector agencies, or the committees that operate some partnerships. Also, the partnership mode of operation has given access to contacts, expertise and knowledge within a wide range of organisations in the region. Hosting the project at the University of Plymouth has helped to establish its credibility and neutrality. In addition, personal factors have undoubtedly been critical: the partnership has been fortunate in having had two particularly able and energetic officers, who have earned the respect and trust of the numerous partner agencies.

Despite occasional differences, the individuals most closely involved with the partnership have collaborated effectively towards common aims.

Some queries and constraints

Despite the generally positive profile that the Devon and Cornwall Rail Partnership has earned, it is possible to raise a number of questions and qualifications. Perhaps the most awkward, yet elusive, query is just how far the initiatives undertaken have contributed *significantly* to an increased use of local branch lines, a reduction in the impact of the private car for rural recreation, or stronger rural development. It would not be excessively cynical to suggest that the project has been fuelled by strong faith that these objectives have indeed been met, in the absence of really convincing evidence (Eaton and Holding 1996). In terms of the full spectrum of rural recreation and tourism in south-west England, private road traffic still appears very dominant, with plenty of indicators that this trend could continue, rather than reverse. The sheer *scale* of change required for longer-term sustainability requires much more profound shifts in public policy and individual preferences and action. The great majority of destinations for leisure trips in the countryside are beyond the reach of the limited rail network, and in many cases, beyond that of the bus network, a pattern that is exacerbated by the diversification that characterises the new rural tourism.

The weaknesses of the partnership approach must also be considered. There are many benefits from the Rail Partnership's location within a complex set of inter-organisational and personal linkages, but these also impose excessive demands on time taken up with attendance at meetings, events and presentations. There is a need to satisfy a varied set of sponsors and their differing demands, objectives and ways of working. The procedures for providing feedback and making claims for budgetary contributions varies between partner organisations.

There are disadvantages associated with relatively short, fixed-term projects: effective strategic planning and action is inhibited (Hutchinson 1994), with real problems in achieving continuity (Countryside Commission 1987). Similarly, relationships with community organisations and businesses take time to mature to the point where there is sufficient understanding and trust. Much energy and time must go into project renewal and convincing partners that they should continue their financial contributions against a background assumption that such projects should become 'self-sustaining'. Here, operational and organisational 'sustainability' become key issues. The partnership has benefited greatly from the participation of the Countryside Commission and the RDC, but both display something of a 'pump-priming culture' in their unwillingness simply to renew support for existing projects, which are expected to become self-sustaining. Renewal requires a project to appear 'different' in terms of objectives and operations, which could imply deviation from earlier priorities and successful projects, and an excessive diffusion of effort in pursuit of innovation. Where initial funding sources are unavailable or expansion seems desirable, alternatives must be pursued. The Devon and Cornwall Rail Partnership has added Plymouth City Council and the Dartmoor National Park Authority as new partners, and has acquired funding from the European Regional Development Fund (5b), and through the Regional Challenge process. However, such success comes at a price of much time, ingenuity and determination spent on the complex application procedures.

Other issues relate more specifically to the partnership's status as a rail-focused tourism initiative. The operation of an economically viable, year-round public transport service may be incompatible with optimising the potential of railways as a leisure experience. For the former, the emphasis is on low-cost, basic transportation, using utilitarian equipment designed to operate on a wider network of non-leisure services. Recreational rail travel often has different priorities and client expectations. The ideal modern 'recreational train' might feature excellent visibility, more comfort and space than is normally offered to commuters, flexible space on board for the sale of refreshments, supply of information or entertainment, a public address system, and access and accommodation for wheelchairs and bicycles. Alternatively, the rail leisure market is strongly attracted to the potent nostalgia of a slower pace of travel and 'railway heritage', including vintage trains and infrastructure that are very inappropriate in modern railway operation.

The relationship between the partnership and the railway industry has also sometimes been ambivalent. Despite strong support by individual managers, on occasions the project has appeared marginal to mainstream railway affairs, and it has rarely been integrated into management decisions and operations at a strategic level. Underlying tensions between the train operator and other partners has sometimes been apparent, with a perception among some sponsors that the partnership has at times been used as a convenient low-cost marketing tool for the railway company, but with insufficient support and influence to fulfil its potential. It would also not be surprising if some in the railway industry viewed the partnership as inexperienced and idealistic interlopers. Such strains highlight a distinct weakness in the fragmented, 'devolved' forms of governance that the partnership represents, and have been reported for other comparable ventures (Wilson 1997).

Future prospects and concluding comments

Despite the observations above, there is a determination to continue and expand the partnership beyond its second three-year term in 1997. The importance of a strategic, coordinating role across the two counties is recognised, especially given the uncertainties and opportunities generated by rail privatisation. As well as marketing and promotion, there is scope for further development and research activity. Examples include progress towards improved accessibility to trains for the less able and for the carriage of cycles, and exploration of the potential of railways as 'corridors' of appropriate development (Mowforth 1996).

Perspectives on the future of the Devon and Cornwall Rail Partnership and similar ventures must be tempered by a degree of realism and uncertainty.

The scale and momentum of car-based recreation appears enormous relative to the possibilities for public transport. Besides the entrenched resistance of the consumer and the tourist industry, there are questions relating to the future status of rural public transport, as well as the priorities and powers of the relevant public sector agencies. The longer-term implications of rail privatisation are still unclear; although a positive view would look for a more innovative and flexible response by the rail industry, there are also possible difficulties. While the current level of services appears to be protected through the terms of the franchise awarded to South Wales and West Railway, the company has committed itself to a reduction in the overall subsidy it receives, from £70.9 million per annum to £38.1 million by 2003–04 (Office of Passenger Rail Franchising 1996). This target can be met only by a large-scale increase in revenue and reduction of costs. This could lead to a loss of commitment to 'marginal' rural services by both the train company and Railtrack (as owners of the infrastructure), which would feel compelled to divert resources to more commercially attractive operations elsewhere.

A more optimistic speculation could present the partnership as playing a role in the transition towards an era when a higher degree of intervention and regulation on behalf of environmental and social sustainability becomes more acceptable. Although still a distant (and contestable) prospect, there may be a case for more deliberate strategic integration of public transport within the planning and development of recreation and tourism, following the principles reflected in PPG13 (DoE 1994b). As a broad illustration, the Tamar Valley line could be projected as the focus of a 'corridor' of appropriate rural recreation, with particular respect to the population of Plymouth. This would require stronger, more 'proactive' guidance and support for recreational development that is accessible from the railway – a move considerably beyond the level of activity to date, which has been based on publicising the rail journey itself and existing tourism resources. However, there are clearly major barriers to such an approach, including problems of funding, and the strong probability of hostility from local residents. Certainly, the spatial concentration of recreation and tourism implied by any switch back to public transport access raises other issues of sustainability. Nevertheless, the next decade could well see the emergence of more radical thinking on policy for rural tourism and its transport needs.

Chapter 12

Market segmentation and ecotourism development on the Lower North Shore of Quebec

John Hull

Since the 1980s, the Lower North Shore of Quebec, a remote region on the northern coast of the Gulf of St Lawrence, has witnessed an increasing number of tourist arrivals (Statistics Canada 1988; Ministère des Transports du Québec 1988; Jones 1991; Jones 1993). The increasing demand for tourism, coupled with the closure of the fishery in the region, has resulted in a new interest on the part of local residents in promoting a small-scale tourism industry. Historically, tourism in the region has largely been controlled by outside tour operators. However, local residents are now examining ways in which to capture a corner of the tourism market. Through local initiatives, natural and cultural resources are being safeguarded, and local receipts and employment opportunities are increasing.

As the global tourism industry becomes increasingly flexible, segmented and customised, remote communities such as those on the Lower North Shore are responding by targeting niche markets (Poon 1994). Ecotourists have become a specific target market of industry planners on the Lower North Shore. They are known to be responsible travellers who are more likely to contribute to local nature conservation and economic development initiatives (Eagles 1992; Western 1993).

Wight (1995b) has identified divisions or subsets within the ecotourism market. She categorises ecotourists as having a range of interests in nature that varies from specialist (scientific) to generalist (casual interest) and who engage in activities that require a high to low degree of physical effort. Visitors to the Lower North Shore have been analysed in the context of these subsets through an analysis of respondent characteristics and travel motivations. Once a profile of visitors is completed their levels of satisfaction and expenditure patterns will assist in forecasting future demand. From these data, it is clear that the sustainability of tourism on the Lower North Shore will ultimately be dependent on how well local planners are able to identify and target specific subsets of the ecotourist market.

The destination

The Lower North Shore extends approximately 380 km from Natashquan to Blanc-Sablon (Figure 12.1). In 1991, the population was approximately 5,200,

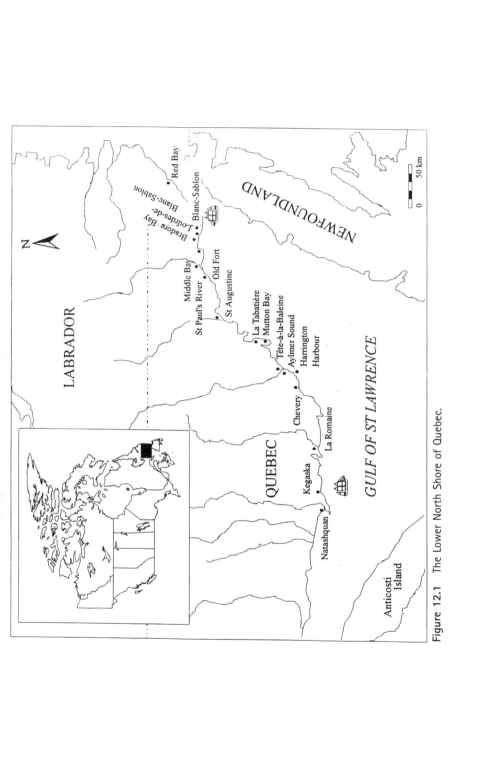

Figure 12.1   The Lower North Shore of Quebec.

spread across fifteen communities. The communities are divided into three political jurisdictions: the municipality of the Lower North Shore, the municipality of Bonne Esperance, and the municipality of Blanc-Sablon (Statistics Canada 1991). Of the total population, 65 percent of the residents are anglophone, 21 percent francophone and 14 percent Montagnais (Blanchard 1994). The Montagnais are a group of Native Americans that have been present in the region since the prehistoric period approximately 9,000 years ago (Mak 1981). Emigration is now a problem with the closure of the fishery. In the period 1986–91, the total population in two of the three municipalities, the Lower North Shore and Blanc-Sablon, decreased by 9.3 and 4.0 percent, respectively (Statistics Canada 1991).

The commercial fishing industry has been the predominant industry since European settlement in the seventeenth century, with over one-third of the local employment in the harvesting and processing of fish (Statistics Canada 1987). At present, the construction industry and crab, shrimp, and scallop fishing provide the majority of jobs. Even so, unemployment levels in the three municipalities are 38, 81.5 and 56.7 percent, respectively (Statistics Canada 1991). Regional and local development agencies view ecotourism as one way to help to diversify the economy.

Ecotourism, defined as responsible travel to natural areas that conserves the environment and improves the welfare of local people (Western 1993), is seen as having potential to contribute to the local economy of the Lower North Shore by providing a source of employment, income and government revenue that will hopefully be sustainable. Sustainable tourism development is defined as tourism that is developed in an area over a period of time so that it does not degrade or alter the environment and prohibit the successful development of other activities and processes (Butler 1993).

The future of ecotourism on the Lower North Shore is ultimately dependent upon the health and viability of the natural resource base, particularly of the marine life found along the coast. The Gulf of St Lawrence is today a world-renowned site for observing marine mammals such as the humpback whale. One of the largest concentrations of whales is found on the Lower North Shore between Blanc-Sablon and Old Fort Bay (Fuchs 1995; Nove 1995). The coast is also home to over 100,000 nesting seabirds scattered across ten migratory seabird sanctuaries. These are some of the oldest sanctuaries in North America, established in 1925 and managed by the Canadian Wildlife Service (Blanchard 1984). If ecotourism is to be successful, the proper management and protection of these wildlife resources is critical to ensuring the development of a viable industry.

Methodology

In the summer of 1995, a survey of departing visitors was conducted on the Lower North Shore at the Kegaska and Blanc-Sablon ferry terminals. These terminals were selected because they target the two major travel circuits available to visitors (See Figure 12.1, page 147). Kegaska is the gateway to the Lower North Shore from the west, being serviced by the *Nordik Express* ferry from Rimouski, Quebec, which continues on to Blanc-Sablon and back once a week. Blanc-Sablon is also served two to three times a day by the *Northern Princess* ferry, which links the island of Newfoundland to the Lower North Shore from St Barbe. Previous tourism studies in the region indicated that the ferries servicing the region were the major means of transport for tourists (Jones 1993). Over a ten-week period during the high tourist season, 538 questionnaires were collected. This sample size represents an estimated 6 percent of total tourist arrivals in 1995.

Respondent characteristics

Historically, tourist profiles have been generated to plan and manage visitor demand at a particular destination. Analysing tourist demand has traditionally been based on one of two approaches: a socio-economic approach or a psychological approach. The socio-economic approach attempts to establish a correlation between a visitor's actions at a particular destination and their social position (Lowyck *et al.* 1992). Mathieson and Wall (1982) argue that visitor attitudes, perceptions and motivations at a destination are influenced by socio-economic characteristics such as age, education, income, residence and family situation. The following section analyses the socio-economic characteristics of visitors to the Lower North Shore to determine if these characteristics reflect segmentation in the visitor market.

Visitors to the Lower North Shore can be characterised as follows. First, visitors surveyed were evenly divided between men and women. Most visitors were highly educated, with approximately 50 percent of the visitors to Blanc-Sablon and 35 percent of the visitors to Kegaska being graduates. Annual household income before tax of visitors to Kegaska and Blanc-Sablon was between C\$50,000 and 59,000. The average age of independent tourists was 50, while tourists on package trips were generally older, with an average age of 58. On performing a z-test, the age difference between independent tourists and package tourists was found to be significant ($z = 18.4$), suggesting two subsets of visitors and a potential need to cater to different age groups. The age of travellers has been shown to have a direct influence on what services are developed in a region (Gunn 1994).

On the Lower North Shore, language and origin of visitors are important considerations in defining the tourist market. In terms of maternal language, visitors to Kegaska were overwhelmingly French-speaking (70 percent) while over 80 percent of visitors to Blanc-Sablon were English-speaking. Bilingual services are being developed to cater to a French-speaking clientele travelling primarily through Kegaska and an English-speaking clientele travelling primarily through Blanc-Sablon.

Three-quarters of the visitors to Kegaska were from Canada, with the majority of the remaining visitors being from France (10 percent) and the United States (10 percent). At Blanc-Sablon, 55 percent of the visitors were from Canada with 40 percent of those sampled from the United States. In analysing the origin of Canadian visitors to the Lower North Shore, 52 percent

Table 12.1 Origin of Canadian visitors (%)

	Kegaska	Blanc-Sablon
Quebec	90.2	15.3
Ontario	9.7	27.9
Maritimes	0	51.4
Prairies	0	1.1
British Columbia	0	3.3
Far North	0	0.7

of the visitors to Blanc-Sablon were from the Maritime Provinces, while in the case of Kegaska, 90 percent were from Quebec (Table 12.1). The origin of visitors suggests that marketing strategies at the present time are attracting a specific European and North American international market and regional Canadian markets.

The socio-economic data gathered at Kegaska and Blanc-Sablon are comparable to data collected by other researchers investigating ecotourism development. In general, ecotourists tended to be older, to have annual incomes above US$40,000, and to be more highly educated than the average traveller (e.g. Eagles 1992). On the Lower North Shore, age, language and origin of visitors are important subsets of the ecotourism market that have been identified. However, socio-economic analyses have often been criticised for not being able to provide a sufficient understanding of consumer behaviour and for not discriminating enough between consumers (Ryel and Grasse 1991; Lowyck *et al.* 1992). In the next section, the psychological motivations of Lower North Shore visitors help to classify ecotourism subsets further through an analysis of visitor behaviour.

Motivations

Psychological approaches classify people into groups according to their lifestyles. Lifestyles are distinctions in people's behaviour that are identified and categorised to distinguish different types of respondent. In a comparative study of Canadian tourists, ecotourists were found to be more motivated by features such as wilderness and parks than the rest of the Canadian population in choosing a destination (Kretchmann and Eagles 1990).

A recent report by the Canadian Tourism Commission (KPMG Consulting 1995) identifies a four-step framework for reviewing tourism marketing activities and is useful for understanding the attraction motivations for ecotourists to the Lower North Shore in 1995. In the initial awareness phase, the traveller is made aware of the potential features and benefits of a particular destination, increasing their interest and curiosity. In the travel planning phase, the tourist contacts an organisation for information on a particular destination. The commitment phase involves the actual booking of travel arrangements to a particular destination, while the experience phase identifies the activities of the tourist at the destination.

The awareness phase

In 1995, there were numerous marketing strategies to attract clients to the Lower North Shore. Magazine and newspaper articles on the North Shore of the Gulf of St Lawrence were featured to promote an interest in the region among readers. The largest French daily newspaper in Quebec, *La Presse*, ran a fourteen-part travel series on the region recounting 'the voyages, the ship-wrecks, the Amerindians, the hunters, the fishermen, the missionaries, and the whales . . . providing an invitation for adventure' (O'Neil 1996). *La Mer & Océan*, a French travel magazine in Europe, and *Photodigest, L'actualité* and *Motoneige* in Canada are magazines that ran feature articles inviting visitors to experience wide open spaces, the small fishing villages, seabird sanctuaries, flora and fauna, and icebergs frequenting the coast. A regional tour promoter also developed web pages on the Internet featuring tourism offerings on the Lower North Shore (Hull 1996).

In analysing the data from the visitor survey, tourists' curiosity and initial reasons for coming to the Lower North Shore clearly varied (Table 12.2). In general, the landscape was selected by all tourists as the most important reason for coming. In Kegaska, culture and whales, while in Blanc-Sablon, icebergs and culture, ranked second and third, respectively. The data reveal that the Lower North Shore at the present time supports tourists interested in both nature and culture.

The travel planning phase

In this second phase, the data reveal that government information, tourist guide-books and friends were the most important sources of information in trip planning (Table 12.3, overleaf). Mathieson and Wall (1982) argue that information on the destination is often transmitted through both formal and informal sources.

The Association Touristique Régionale (ATR), the regional tourist association for the Lower North Shore, in cooperation with Tourism Quebec, distributes

Table 12.2   Initial reasons for coming to the Lower North Shore

Kegaska	Rank	Blanc-Sablon	Rank
landscape	1	landscape	1
culture	2	icebergs	2
whales	3	culture	3
adventure	4	residents	4
residents	5	adventure	5
icebergs	6	relaxation	6
relaxation	6	whales	7
seabirds	8	fishing	8
hiking	9	other	9
arts/crafts	10	hiking	10

Table 12.3   Sources of information used in planning trip

Kegaska	Rank	Blanc-Sablon	Rank
tourist guide	1	friends	1
friends	2	government	2
government	3	tourist guide	3
newspaper	4	magazine	4
relatives	5	relatives	5
magazine	6	other	6
television	6	television	7
other	8	travel agent	8
travel agent	9	newspaper	9
tour operator	10	tour operator	10

a tourist guidebook that is updated once every three years for tourists travelling in the region. The ATR office in Sept-Îles serves as the formal contact for obtaining tourist information on the region. However, a number of tourists complained that tourist information was not readily available on the Lower North Shore. A visitor from Newfoundland commented: 'The Lower North Shore needs more explanation on the history and culture of the area for tourists. What happened there is unknown to the tourist and not readily available locally. There seems to be very little information put out by the Quebec Government about this area.'

Steps are being taken to ameliorate this problem. Three local tour operators developed brochures in 1995 to advertise their businesses, and the Quebec–Labrador Foundation published a visitors' guide featuring the natural and cultural history in the Blanc-Sablon region and is planning a series of visitor guides for the coast.

Numerous visitors to Blanc-Sablon also commented that friends and acquaintances had recommended that they visit the region. Travel information from friends has been found to be one of the most trustworthy sources of information for visitors (Walker 1995; Withiam 1996). Previous visitors to the Lower North Shore, therefore, play an important role in influencing tourist demand. Almost 80 percent of those surveyed in Kegaska said that they would definitely recommend the Lower North Shore as a tourist destination to friends. Three-quarters of those surveyed in Blanc-Sablon said that they would definitely or probably recommend the Lower North Shore to friends.

Informal sources play an important role in increasing the awareness of potential visitors, educating visitors on what their experience will be like, and describing the destination resources and attractions. They also play a role in attracting a certain type of tourist. One visitor from Quebec commented that they would recommend the Lower North Shore 'to friends of nature, who are adventurous'. A sustainable tourism industry will depend upon attracting such conservation-minded tourists. Tourism planners on the Lower North Shore are attempting to identify and target an ecotourist market that travels consciously.

The commitment phase
The data collected on pre-trip planning reveal that over 80 percent of those surveyed visited the Lower North Shore for vacation, with between 70 and 80 percent of visitors coming to the Lower North Shore for the first time. While a majority of tourists are first-time visitors, there is still a need to cater to repeat visitors. One visitor from the Maritime Provinces commented: 'this is our tenth visit in as many years'.

In terms of the visitors' means of travel to the Lower North Shore, over 60 percent of those surveyed from Kegaska travelled on a package tour, while in Blanc-Sablon, over 80 percent of those surveyed were free and independent travellers (FIT). Only 44 percent of visitors to the Blanc-Sablon region made their 1995 summer travel plans prior to May of that year (Government of Newfoundland and Labrador 1996). The majority of visitors planned their visit during the summer. These numbers indicate that marketing of the region must be sensitive to both the tourist who plans their trip four to five months in advance and those who might make a more spontaneous decision. A visitor from Ontario commented: 'We had no plan to visit the Lower North Shore/Labrador Straits when we left on our family holiday to Newfoundland. It was an addition to our trip based on recommendations we received en route. We appreciated the lack of commercialism and would hate to see that change.'

There were numerous visitors to Newfoundland who did not hear of the attractions in the Blanc-Sablon region until after they had arrived on the Northern Peninsula of the island of Newfoundland. Visitors to Kegaska also commented that they were not aware of the boat trip to Blanc-Sablon until after they had arrived in Sept-Îles and visited the Association Touristique Régionale. Flexible tourist services that cater to the package tourist and the free and independent traveller are currently being developed.

The experience phase
While on the Lower North Shore, visitors surveyed in Kegaska indicated overwhelmingly that their main destination was Blanc-Sablon (65 percent). This is largely due to the last stop being Blanc-Sablon for the *Nordik Express* ferry on its trip from Rimouski. For visitors surveyed at Blanc-Sablon, over 70 percent indicated that Red Bay in Labrador was their main destination due to the Basque whaling site. These data reinforce the fact that there are two main travel circuits on the Lower North Shore at the present time due to ferry services in the region.

The ferry services also influence the amount of time visitors are spending on the Lower North Shore. In Kegaska, over 90 percent of the visitors surveyed spent four or more days, while in Blanc-Sablon over 80 percent of the visitors surveyed spent one day or less. Visitors travelling through Kegaska are restricted to a trip of at least four days when travelling by ferry. Visitors to Blanc-Sablon are connected by road and a ferry to Newfoundland that runs approximately two to three times a day in the summer months.

In terms of the activities that visitors engaged in on the Lower North Shore, the data from Blanc-Sablon and Kegaska reveal that there is a profound

Table 12.4 Activities visitors engaged in while on Lower North Shore

Kegaska	Rank	Blanc-Sablon	Rank
nature study	1	visiting historic sites	1
taking photos	1	visiting parks	2
meeting locals	3	meet locals	3
boat trips	4	taking photos	4
visiting historic sites	5	purchasing arts/crafts	5
bird-watching	6	nature study	6
hiking	7	camping	7
visiting parks	8	fishing	8
fishing	9	bird-watching	9
purchasing arts/crafts	10	hiking	10
camping	11	boat trips	11

interest in the natural and cultural heritage of the region (Table 12.4). In Kegaska, a cumulative ranking of visitor activities reveals that nature study, taking photos, meeting locals and riding the ferry were the top four activities ranked by respondents. In Blanc-Sablon, visiting historic sites, visiting parks, meeting locals and taking photos were the top four activities. A visitor's comments from Jackson, New Hampshire, illustrate the responsibility that local residents have in protecting the fragility and value of the natural and cultural attractions in the region: 'Try to keep the natural and cultural history alive. This is a beautiful, unspoiled place. Development brings highways and Wal-marts. They ruin the environment and the local culture.'

The data suggest that Kegaska and Blanc-Sablon are attracting a generalist clientele that engages in activities that require a low degree of physical effort. However, visitors to the coast differ in their needs and services. Blanc-Sablon is attracting greater numbers of free and independent tourists who stay for a short time, travel by automobiles and campers, and are interested in cultural sites, while Kegaska is catering to a package tourist interested in staying for longer periods and engaging in nature study while on a boat cruise to remote locales in the region. The next two sections analyse the satisfaction levels and expenditure patterns of these visitors to assess the potential sustainability of tourism in the region.

Levels of satisfaction

Levels of satisfaction play an important role in influencing local tourist expenditure and future tourist demand through informal referrals. In general, visitors surveyed were satisfied with their trip to the Lower North Shore, with the overall satisfaction levels for independent and package tourists at 70 percent. Satisfaction levels showed that the natural scenery and the friendliness of residents ranked first and second, respectively (Table 12.5).

However, a number of factors differed between the two types of traveller. Package tourists were more satisfied with opportunities for wildlife viewing,

Table 12.5   Satisfaction level of tours

	% Satisfaction independent	% Satisfaction package tour
Cost of arts/crafts	61.41	58.64
Availability of arts/crafts	65.24	55.60
Availability of local food	65.80	66.92
Cost of food/drink	66.71	68.55
Quality of arts/crafts	68.01	62.10
Cost of travel within region	69.34	65.22
Cost of travel to region	70.23	72.73
Availability of information	71.77	69.44
Availability of lodging	72.23	74.02
Wildlife viewing	72.34	78.03
Ease of travel within region	72.92	61.00
Cost of lodging	73.62	72.16
Cultural sites	78.32	74.18
Quality of lodging	79.60	78.18
Opportunities to meet locals	81.59	78.42
Friendliness of residents	93.90	92.00
Natural scenery	94.78	92.81

ranking it higher than independent tourists. This factor is a reflection that a majority of package tourists surveyed were travelling on the *Nordik Express* ferry and, therefore, were physically on the Lower North Shore for a longer period of time. Local guides in a number of villages also offered short excursions while the *Nordik Express* was in port. As a result, these visitors visited more remote locales where it was possible to see wildlife.

Second, ease of travel in the region received a higher satisfaction ranking from independent tourists than from package tourists. This is a result of the strict schedule of the *Nordik Express*, which limited tourist visits in the villages often to little more than 90 minutes. The majority of independent tourists travelling through Blanc-Sablon have easy road access to the communities and attractions.

Finally, both groups ranked the quality, availability and costs of arts and crafts in the region the lowest. Local handicrafts have a tremendous sales potential and even though individual transactions may be small, their total sales can be significant to the local and even national economy (Healy 1994). Travellers interested in culture and nature, such as those on the Lower North Shore of Quebec, have been shown to have an interest in local crafts and are potential consumers of these goods (Eagles 1992; Healy 1994). Handicrafts provide employment and income locally, have potential to revitalise local craft traditions, and 'offer the possibility of using local materials sustainably' (Healy 1994).

A number of communities and organisations are taking steps to revitalise local craft traditions and heighten their visibility. In Tête-à-la-Baleine, Le Tannerie Abri-Nor employs a local resident to produce hats, gloves and stuffed animals. The Federal Office of Regional Development in Sept-Îles is also currently working on a number of projects to improve local cottage industries. One specific project in Natashquan is closely tied to the tradition of berry picking. In the fall of 1996, local residents were employed in the production of a red berry liqueur for sale at regional and provincial liquor stores. The Coaster's Association, a regional non-profit organisation working to promote and support the ideas and actions that contribute to the vitality of the Lower North Shore, has also completed a community directory available to visitors to publicise tourist services in the region (Gascon 1994). With only 8.3 percent of the independent tourists and 9.8 percent of the package tourists surveyed purchasing local handicrafts, there is tremendous potential for improving the development of cottage industries.

Patterns of expenditure

Researchers disagree on the economic impact of nature tourists on local communities. On the one hand, there is the argument that since these visitors spend most of their time out on the land or in the wilderness their economic impacts on local communities are minimal (Wall 1993). However, Boo (1990) found that nature-oriented tourists had higher daily expenditures than those tourists who were not nature-oriented. Grekin and Milne (1996) also argue that ecotourism is an industry where the physical isolation of a destination may work to its economic advantage by providing a taste of the unknown and the untouched. Stoffle *et al.* (1979), in a study conducted in the southwestern United States, also found that tourists who felt positive about residents at a particular destination were likely to purchase items to remember their experience. With a majority of visitors satisfied with the friendliness of residents on the Lower North Shore, the potential for increasing the economic impacts of tourism in the region through hospitality services is great. By understanding how tourists are spending their money at present, planners will be better able not only to continue to target services to meet visitor needs, but also to continue to develop marketing strategies that attract them. In examining the present average daily expenditure patterns of visitors, it was found that package tourists, at C$70.43, had a higher average daily expenditure than independent tourists (C$49.90) (Table 12.6). Accommodation was the largest expenditure, with package tourists spending on average C$42.04 and independent tourists spending C$11.76. For package tourists, accommodation costs represented 59.6 percent of their average daily expenditure, while for independent tourists accommodation costs represented only 23.8 percent. Package tourists' second-largest expenditure category was transport at 22.2 percent, while for independent tourists meals were the second-largest category at approximately 18 percent. Expenditure patterns show that over 82 percent of the package tourists' costs are restricted to accommodation and transport, while independent tourists, even though they spend less overall, are spending more money in different sectors of the local economy and contributing more to the sustainability of the industry.

Table 12.6   Average daily expenditure for FIT/package tourists

Expenditure category	FIT average daily expenditure C$	%	Package average daily expenditure C$	%
Accommodation	11.76	23.8	42.04	59.6
Fuel	6.98	14.1	3.31	4.7
Meals	8.80	17.8	6.34	9.0
Transport/tours	6.65	13.4	15.66	22.2
Museum/park fees	1.07	2.1	0.49	0.7
Groceries/supplies	5.98	12.1	1.62	2.3
Arts/crafts	6.66	13.4	0.62	0.8
Other	1.50	3.0	0.35	0.4
Total	49.40	100	70.43	100

The goal of industry planners is to increase opportunities for tourists to spend money locally in order to generate local revenue and employment. One of the major problems historically is that package tourists to the region travel with tour companies from off the coast. Most of their costs are paid up-front to the tour operator, with little of the money being spent on the Lower North Shore. With three tour operators from on the coast offering package trips in 1995, local economic impacts should increase.

One of the major criticisms of package tours that are externally controlled is that a large percentage of the revenues that are generated, in some cases as much as 55 percent, never reach the destination (Boo 1990; Alderman 1992). Reducing economic leakage is therefore an important consideration that local communities must address in the successful development of the industry. Leakage is not a predetermined outcome but depends on the choices of both the tourist (e.g. staying in locally run accommodation and buying locally produced products) and the tour operator (e.g. maximising use of local materials, people and products) (Laarman 1987). Therefore, successful marketing campaigns on the Lower North Shore that emphasise local services and the participation and development of small-scale, locally run businesses, that do not require major capital investments, are critical for a healthy local economy (Alderman 1992).

Conclusions

The impacts of tourism development at a particular destination are determined by what type of tourist is attracted to the region (Milne and Wenzel 1991). In analysing socio-economic and psychological factors of visitors to the Lower North Shore in 1995, it was revealed that visitors to the coast can be categorised as ecotourists: 72 percent of the visitors surveyed to Kegaska and 64 percent of the visitors to Blanc-Sablon considered themselves ecotourists. Socio-economic data revealed that age, language and origin of visitors are important subsets of the ecotourist market that should be addressed in developing future marketing strategies.

Psychological factors revealed that tourists are curious to learn more about nature and culture in the region. Word of mouth is also an important source of information in trip planning to the region. Visitor activities are of a generalist nature and require a low degree of physical effort. Photography, nature study, bird watching, visiting cultural sites and meeting locals are the primary activities most ecotourists engage in on the Lower North Shore. However, these activities vary depending upon which tourist circuit is taken in the region.

Satisfaction levels suggest that the sustainability of tourism will be enhanced by offering more opportunities for wildlife viewing in Blanc-Sablon, through improving the ease of travel through Kegaska, and in developing local cottage industries all along the coast that make use of local materials, people and products to improve local profits and reduce revenue leakage. With the number of tourists engaged in ecotourist activities estimated to be expanding by approximately 20 percent a year worldwide, it is essential that steps are taken to minimise ecological and cultural impacts so that the industry remains economically viable (Ziffer 1989; Lindberg 1991).

At a tourism conference in 1995 on the Lower North Shore administered by the Coaster's Association, a tourist mission statement was adopted by the 75 participants that reinforces the commitment of local residents to promoting a sustainable industry. The goals from the conference were to

• preserve the natural resources and attractions;
• respect the people, traditions, history and unique way of life; and
• promote local employment and economic diversification.

The newly formed Lower North Shore Tourism Development Corporation, with representatives from the three cultures on the coast, has been mandated to implement the tourist mission for the region. One of their first projects is the development of a regional tourism guidebook that specialises in attracting specific subsets of the ecotourist market. This guidebook, along with the continued hard work, participation and hospitality of local residents, should continue to make the Lower North Shore an attractive ecotourist destination.

Acknowledgements

I would like to thank Simon Milne and the McGill Tourism Research Group at McGill University for assistance in preparing this manuscript for publication. I would also like to thank the Quebec Ministry of Education, the Province of Quebec's Society for the Protection of Birds, the Quebec–Labrador Foundation and McGill University for financial support in conducting the research.

Chapter 13

Development in Nepal: the Annapurna Conservation Area Project

Matt Pobocik and Christine Butalla

This chapter examines the Annapurna Conservation Area Project (ACAP) and the effects of trekking tourism in the Annapurna region of Nepal. Tourism-related environmental problems such as solid waste disposal, pollution and resource consumption in Nepal are especially severe. Water, land and air pollution are all involved. The ACAP was selected as a case study because it was founded on ecotourism principles and has been in operation for ten years. Ecotourism is a form of tourism that operates on the principle that environmental damage should be minimised, while economic benefits are maximised. The main benefit of ecotourism is that it has the potential to provide needed capital for the local and national economies without exceeding ecological or cultural carrying capacities. This study focuses on economic and environmental impacts of ecotourism, with a primary focus on the consumption of fuelwood by trekking tourism. Fuelwood consumption is an important issue because the forest supplies most of the energy needs in Nepal.

Nepal is home to eight of the highest mountains in the world, including Mount Everest, making it a primary destination for mountain trekkers. Trekking now accounts for a quarter of Nepal's tourism, which has increased from 50,000 in 1970 to 300,000 in 1993 (HMGN 1994). Sagarmatha National Park (the region around Mount Everest) and the ACAP are the two most popular trekking regions in Nepal (Figure 13.1, overleaf). The ACAP, which receives roughly 60 percent of all trekkers in the country (HMGN 1994), can be divided into three major management areas: (1) the Upper Mustang Conservation and Development Area, which was added in July 1992; (2) the Annapurna Conservation Area, which comprised the original boundary of the ACAP; and (3) the Annapurna Sanctuary, which is a sacred valley within the Annapurna Conservation Area (Figure 13.2, page 161).

The ACAP was created in order to alleviate environmental degradation linked to trekking tourism by stressing conservation and development. The goals of the ACAP are to involve locals in the management of the reserve, to provide fees paid by trekkers directly to local inhabitants for management of the reserve, to provide economic benefits to locals, and to preserve the environment in Nepal's Annapurna Conservation Area Project (Harrison 1992; Wells 1994). The ACAP's Minimum Impact Code stresses that tourists should

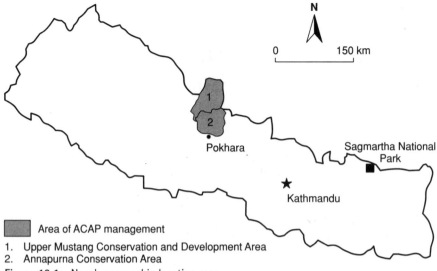

Figure 13.1   Nepal: geographic location map.

conserve firewood, stop pollution and be guests (KMTNC 1994a). But, is ecotourism as practised in Nepal really ecotourism? Three issues are explored in this research: the economic benefits and environmental impacts of group and independent trekking; tourist environmental attitudes; and tourism policy. The environmental impacts of group and independent trekking are discussed in terms of fuelwood use.

Methods

This study was conducted in Nepal from July 1994 to May 1995 supported by grants from the National Security Education Program and Miami University. The use of participant observation and social survey methods allowed the researcher to gather both quantitative and qualitative data about the economic and environmental impacts of trekking and to determine tourist environmental attitudes, economic impacts of tourism, and fuelwood consumption. Tatopani was selected as a village study site because it is located on the most popular trekking route in the ACAP. Trekkers, local residents, trekking tour guides, travel agents, lodge owners, ACAP personnel, and government and NGO officials were interviewed.

Results

The two major types of trekkers in Nepal are group and independent trekkers. Group trekkers are those who are participating in an agency-arranged trek and are camping, while independent trekkers are those who did not use a trekking agency and stay in lodges.

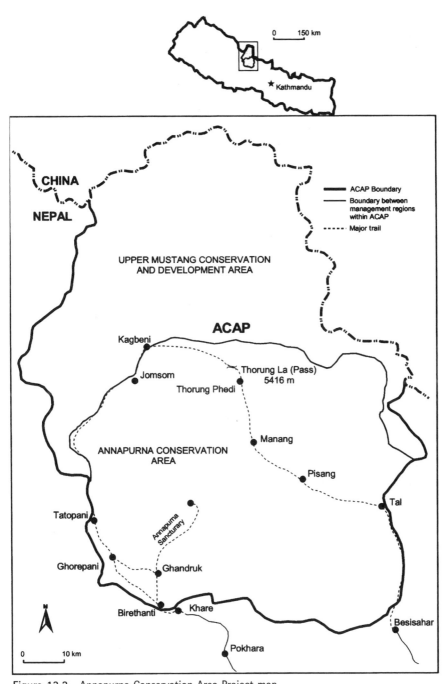

Figure 13.2   Annapurna Conservation Area Project map.
Source: Gurung and De Coursey 1994: 182.

In Nepal, group trekking offers a self-contained camping experience, and in the ACAP the average group size is 10.5 tourists. This self-contained atmosphere lessens interaction with local people, because support staff interact with locals whenever necessary. Ashok Baibya, owner of a travellers' service in Kathmandu, explained that trekking agencies try to insure that members of a group are from similar backgrounds and nationalities to ensure group cohesion. However, this group cohesion tends to isolate tourists, minimising interaction with tourists from other nations and with the people of Nepal. Survey results indicate that 92 percent of the 106 group trekkers and 56 percent of the 385 independent trekkers in the ACAP indicated that they came to Nepal to experience both the people of Nepal and the natural environment. In reality, group trekkers missed much of the possible interaction with the local population and other international tourists, thereby lessening the experience.

Independent trekkers stay in lodges, eat at tea houses (local restaurants), and typically carry a backpack containing necessary goods for trekking weighing between 10 and 20 kg. In the ACAP, the average independent trekker hikes in a group of 2.7 people. The smaller number of trekkers hiking together increases the interaction with locals and other tourists, especially in lodges and eating establishments.

Economic and environmental impacts

In 1993, the ACAP received 39,764 trekkers (HMGN 1994), 72 percent of whom were independent. The large percentage of independent trekkers in the ACAP can be explained by the relative ease of access. Trekkers can reach any major trail head in the ACAP within hours by road from Pokhara. The only airstrip in the ACAP is located in Jomsom, with flights to Kathmandu and Pokhara.

In 1994, each trekker in the Annapurna Conservation Area paid an entrance fee equivalent to US$13 to the ACAP, a total of US$520,000 for the year. In addition to the entrance fee, there is also a US$5 per week government trekking fee for the first four weeks and US$10 for each additional week, providing the government of Nepal with US$600,000 annually in trekking fees. Therefore, the ACAP provides Nepal with over a million dollars annually in fees from trekkers alone. However, the economic impact of trekking tourist dollars reaches far beyond their fees paid to the government and the ACAP.

In the survey, the economic impacts of trekking in the ACAP were broken down by tourist expenditures. Excluding airfare, group trekkers spent an average of US$78 per day in Nepal, while independent trekkers spent US$6.50 per day. These figures were derived from questions concerning average cost per day while trekking, length of trip, total cost of trip, airline used, and nationality of the tourist. Independent trekkers were able to supply expenditures per day while trekking. Group trekkers often could provide only total trekking expenditures or total trip expenditures as most group packages were inclusive. Typically, 60 percent of the cost of group trekking never left the

Table 13.1 Number of guides and porters per person

	Guides	Porters
Groups in tents	0.51	3.1
Independents in lodges	0.07	0.09

Table 13.2 Length of stay and duration of trek

	Days in Nepal	Days trekking
Groups in tents	24.7	19.6
Independents in lodges	42.7	15.8

country of origin. This reduces Nepal's average benefits to US$31 per person per day from group expenditures.

Even at US$31 per day, group trekkers still spent at four times the rate that independent trekkers did in the ACAP. Independent trekkers spent money directly along the trekking route on lodging, food and drink, and souvenirs, while group trekkers' funds that reach Nepal go to agencies in Nepal's regional centres of Kathmandu or Pokhara. The agency in turn buys food and equipment, and hires staff, which contributes to the national economy of Nepal but does little for the local economies within the ACAP. Almost all staff employed by agencies are from regions outside the area being visited. Group trekkers employ more personnel than individual trekkers. On average, group trekkers employ 3.6 support staff per trekker (Table 13.1), which includes guides, porters and cooking personnel. Independent trekkers also hire support staff in the form of guides and porters but at the lower rate of one staff member per ten trekkers. It should also be noted that independent trekkers will often hire staff at the trail head or along the trekking route, thereby employing a higher percentage of locals at a higher rate of pay. Group trekkers camping alongside lodges often pay a small camping fee and do make occasional purchases of drinks or souvenirs. However, group trekkers contribute little to local economies, which is a fundamental factor in the successful trekking agency management paradigm of supplying all needs and reaping all profits. This practice is in direct conflict with the accepted ecotourism paradigm of maximising local economic benefits.

Group trekkers hiked for approximately four days longer in the ACAP than did independent trekkers, but independents spent on average 18 days (75 percent) longer in Nepal than group trekkers (Table 13.2). When considering contributions to the national economy of Nepal, group trekkers, who are spending four times as much per day while in Nepal, on an individual basis still outspend independent trekkers, who stay longer. However, independent trekkers outnumber group trekkers three to one in the ACAP, so in the long run ACAP independent trekkers are contributing more to the national economy than ACAP group trekkers. Taken individually, these data indicate that group trekkers contribute more per person to the national economy than independent trekkers, while independent trekkers contribute more to the local economies.

Fuelwood

Fuelwood use is a key environmental factor within all of Nepal, and the ACAP is aware of problems associated with current fuelwood use. The village of Tatopani was used to determine typical fuelwood usage rates of locals, independent trekkers and group trekkers.

Tatopani fuelwood usage rates

Local fuelwood use was determined by interviewing 97 households with an average size of 4.8 people per household. This comprised 53 percent of the 182 total households in the Tatopani village development district, which is a political jurisdiction in Nepal, reported in *Nepal District Profiles* (Deepak *et al.* 1994). From the household questionnaire, it was determined that it now takes on average one full day to collect one bundle of fuelwood, and that one bundle of fuelwood will last an average of 3.2 days. This household usage rate multiplied by persons per household means that one bundle of fuelwood will supply 15.4 locals per day. Expanding this usage rate over one year provides a total of 19,980 bundles of fuelwood consumed annually for local use by the village of Tatopani.

Staff of Tatopani tourist lodges were also interviewed to determine tourist fuelwood use. Using an average in-season and off-season usage rate for lodges, one bundle will supply 8.6 tourists per day, or 6,570 bundles annually. Using these rates, it can be calculated that tourists are responsible for 25 percent of the total 26,550 bundles of fuelwood consumed in the area.

Three factors must be considered in the calculation of fuelwood usage rates for tourists. First, Tatopani is at a lower elevation and thereby warmer than many villages in the ACAP, so fuelwood use is lower. Second, the Tatopani hot springs supply tourists with hot water for bathing, again reducing possible fuelwood consumption. Third, lodge owners have a bias in reporting fuelwood usage. From interviewing cooks and through direct observation the author concluded that fuelwood usage rates were higher than reported. All these factors reduce the reported tourist fuelwood usage rate to well below the likely rate in the ACAP. Consequently, fuelwood usage rates reported in this chapter for trekkers are conservative.

Support staff use of fuelwood is calculated at local usage rates. These rates are also conservative, because support staff often lack proper clothing and shelter from the weather. Trekking support staff build open fires for warmth and cooking, typically consuming more fuel than local households. The ACAP suggests that trekkers supply their staff with warm clothing to reduce fuel consumption. This is often done, but it is also common for staff to sell the clothing or give it to a relative at home, which can be a substantial benefit to low-wage staff.

Villagers' perceptions

A Tatopani village survey was conducted of 97 households (53 percent of total) to determine village fuelwood use and perceptions of the forest and the ACAP. The results of the village household survey were divided into three categories by walking time from the main trail. The first category, valley, is within one hour from the trail, which is located in the river valley. The second category, middle, is a one- to three-hour walk away from the trail, and the third category, high, is three to five hours from the trail. As the distance from the trail increases, so does the elevation. With this geographical separation, it was expected that there would be differences in locals' perception of tourism and fuelwood use. Excluded from these results are responses from the five households that run lodges.

Differences in local perception of tourism are illustrated in Table 13.3. Only one respondent (11 percent) in the high elevations derived some economic benefit from tourism in the valley below, while at the same time eight (89 percent) respondents indicated that the forest has become worse since tourism. The results found at the middle elevations were that 14 households (34 percent) derived some income from trekking tourism, while 35 households (85 percent) indicated that the forest is worse since tourism. Valley results were that 23 households (49 percent) derived income from tourism and 24 households (51 percent) indicated that the forest has become worse since tourism. This geographical separation, represented by walking time from the tourist trail, is also reflected in the remainder of the responses. Both middle- and high-elevation respondents indicated little benefit from trekking, while the benefits in the valley were more direct.

Table 13.3 Tatopani village survey results

	Valley n = 47 (%)	Middle n = 41 (%)	High n = 9 (%)	Total n = 97 (%)
Life better	47	7	0	26
Life worse	0	0	0	0
Tourist employment	17	0	0	8
Sell to tourist	30	5	0	16
Sell indirectly	49	34	11	39
Electric lights	62	0	0	30
Electric cooking	21	0	0	10
Kerosene cooking	28	0	0	13
Tourist problems	9	0	0	4
Forest better	6	2	0	4
Forest worse	51	85	89	69
Aware of ACAP	9	0	0	4

Valley results indicate that eight households (17 percent) had one or more members employed directly in trekking tourism. This employment was not limited to Tatopani, as some household members are employed as guides or porters, or work in the trekking tourism sector in Kathmandu or Pokhara. Electricity is recent in Tatopani, and 29 valley households (62 percent) have electric lights. Ten of these households (21 percent) use electricity for cooking, which at the date of the interview was limited to electric rice cookers. Four of the village households (10 percent) indicated that tourists were causing problems. Problems indicated by villagers included a loss of culture and inflation due to tourism.

These results point out that 25 percent of the total households think that life is better since tourism came to Tatopani and 38 percent of households derive income from tourism. At the same time, none of those surveyed indicated that life is worse since tourism began in earnest in the 1960s. This can be explained by rural villagers' lack of connection between tourism and their lives. Repeatedly, survey respondents told this researcher to ask these questions of lodge owners because they had nothing to do with tourism. However, 67 percent of households believe that the forest is worse since tourism arrived in Tatopani. Because of the importance of the forest to rural households, this means that most people in the Tatopani Village Development Committee are being harmed by depleting forests without receiving benefits from tourism.

As mentioned above, the average collection time for a bundle of wood is one full day, and each bundle lasts on average 3.2 days. Therefore, if a household contains 4.8 individuals of working age one person must work a full day every 3.2 days to collect fuelwood. If the household is barely able to meet its needs, the collection of fuelwood, while necessary for survival, may also interfere with planting, harvesting or other necessities of survival. This is further exasperated by the need to care for very young and very old household members.

It was reported in the village survey that before trekkers arrived three to four bundles of fuelwood could be easily collected in one day. However, there has also been an increase in the local population in the Tatopani area, in part due to tourism-related employment. Fuelwood use would increase with a population increase, so it should be expected that collection time would also increase. Adding to the environmental strain is the consumption of at least 25 percent of the fuelwood from the area by tourists. Therefore, tourists are contributing significantly to the deforestation of the area, which places a burden on most of the population.

It is also telling that the results of this survey show that only four households in the Tatopani survey were aware of the ACAP. Yet the ACAP has greatly expanded its area of responsibility, adding the Upper Mustang Conservation and Development Area (see Figure 13.2, page 161). Ghandruk is where the ACAP has concentrated its efforts and is a showcase for its management efforts. However, the people of Tatopani, which is only one or two days' walk from Ghandruk, are not yet aware of the ACAP and its activities. The results from the Tatopani village survey indicate that most villagers' standard of living has not improved since tourism.

Trekker survey

The survey results indicated that the majority of trekkers in the ACAP originated from European and North American countries. In general, group treks had higher percentages of older male trekkers and independent treks had higher percentages of younger females. Over 85 percent of both group and independent trekkers were first-time visitors to Nepal.

The trekkers' response to the survey showed environmental awareness and willingness to pay for environmental protection. Over 80 percent of the 533 trekker respondents in the ACAP and 161 trekker respondents in the Everest region indicated that pollution was an environmental problem caused by trekking (Table 13.4). The second most common response to this question concerned the use and overuse of fuelwood. More than 80 percent of all trekkers also indicated that they were willing to pay more to protect the environment while trekking. In the ACAP, of the 386 independent trekker respondents, 90 percent indicated a willingness to pay more to protect the environment. The difference in this percentage may be explained by the number of trekkers who were in groups and the number who were independent trekkers. Agencies handled 55 percent of trekking permits issued to the Everest region in 1993, while only 28 percent were agency-handled in the Annapurna region (HMGN 1994). The higher percentage of group trekkers affected the survey results. Independent trekkers are more willing to increase expenditures (see Table 13.4) for environmental concerns since they are spending less overall. Many of the group trekkers surveyed answered no to this question because they felt that, at roughly US$78 per day, they had already paid for environmental protection. Indeed, many of these group treks were advertised as environmentally friendly. It should also be noted that many trekkers were willing to pay more for the environment only if the money actually went into environmental protection.

The breakdown between group and independent trekkers in the ACAP is further examined in Table 13.5 (overleaf). In the categories of pollution, fuelwood and ACAP awareness the similarity in percentage of responses shows that group and independent trekkers have roughly the same awareness of environmental impacts of trekking and awareness of the ACAP. Again the difference in willingness to pay more to protect the environment can be

Table 13.4 Trekker awareness of environmental problems and willingness to pay more to protect the environment, by region

	ACAP	Everest
Environmental problems caused by trekking		
● Pollution	86%	83%
● Fuelwood use	76%	82%
Willingness to pay more	84%	81%

Table 13.5   ACAP trekker awareness of environmental problems, ACAP, and willingness to pay more to protect the environment

	Pollution	Fuelwood	ACAP	Pay more
Groups in tents	80%	82%	64%	71%
Independents in lodges	88%	77%	65%	90%

explained in that many group trekkers expected the trekking agency involved to maintain environmental protection.

Discussion

Results of this research do not support many of the reports in the popular literature that place the blame for environmental degradation on tourists. The causes and solutions are more complex. According to *The Economist* (1993), 'In Annapurna, the annual number of trekkers now equals the local population of 40,000. They leave litter and smellier souvenirs, and the yaks which carry their supplies over-graze mountain pastures.' Larsen (1993) states that because large group treks are easier to monitor and regulate than independent trekking, group treks contribute less to environmental degradation. In contrast, Shackley (1994) points out that villages receive minimal economic benefit from large group treks. Therefore, local support for independent trekking is stronger than for group trekking because the economic benefits are greater (Zurick 1992; Larsen 1993). It is commonly believed that independent trekking places a greater burden on natural resources than group treks that carry their own fuel and sleep in camps, because extra lodges are built and fuelwood is consumed in the process of providing for independent trekking. The current study refutes these assumptions.

Nepal has been relatively successful at utilising the tourist potential offered by its magnificent landscape; however, this wealth is not distributed evenly. An important point contained in the issue of the extent of benefits is the question of who is benefiting. It appears from the current research that the nation is benefiting from tourism, as are the regional centres of Kathmandu and Pokhara. The problem is that much of the income generated by tourism is very likely to bypass the local communities. This pattern has been observed by others as well. 'Although a major aim of development aid is to provide resources to poor villagers, little development money reaches the village and much of the money is used in projects which help the wealthier groups of the society, much of it is directly stolen' (Metz 1991). Lodge owners are clearly benefiting (Wells 1994). What most local people see is inflation and shortages (Whelan 1991; Nicholson-Lord 1994; Shackley 1994).

The Annapurna Conservation Area Project was created in 1986 to address the needs of human development, nature conservation and tourism management. Instead of a national parks approach, the ACAP used a new approach that allows local populations to remain within and around the conservation area and in essence become the custodians of the resources (Gurung and De Coursey 1994; KMTNC 1994). The Annapurna region never experienced overpopulation, which contributed to massive deforestation in other regions of Nepal. However, the majority of Annapurna's residents are at or below subsistence levels and are over 90 percent dependent upon the forest for energy needs (Gurung and De Coursey 1994). The strain on resources has been from the explosion in trekking tourism. A typical group trek around the Annapurna circuit involves 50 support staff for twelve clients. Fuelwood used by tourists and the construction of more than 700 tea shops have contributed to deforestation in the region (*ibid.*). Almost all food and household items come from outside the region. Therefore, tourism causes local inflation by driving up prices without creating local economic opportunities (KMTNC 1994).

According to Gurung and De Coursey, 'The sustainability of conservation and development projects depends on the participation and motivation of the local people' (1994: 183). The ACAP believes that programme ownership is essential, because the only sustainable motivation for locals is the realisation that they are the principal beneficiaries of tourism. The ACAP provides services to locals and sees itself as a catalyst between locals, outside aid and government agencies (*ibid.*; KMTNC 1994). The ACAP is trying hard to fulfil this mission, yet at the practical level it is not often working. Some of the problems relate to differences in cultural context and how funds are used. Further, ACAP personnel claim that a lack of funding and staff is the reason that much of the original Annapurna Conservation Area has yet to feel the ACAP's presence. Tatopani is a prime example of this lack of presence. There is no visible impact of the ACAP in Tatopani, and only four households responding to the village survey even knew about it.

Tatopani is also a good example of how the benefits of trekking tourism decrease as the distance from the trekking trail increases. Policies that spread the narrow benefits of trekking tourism would help. This can be done by opening new trails into more villages for trekking tourists. At the present time, tourists are not allowed to stray from designated trails. The ACAP's official response to this suggestion was that it is starting a few new trails but wants to improve trekking tourism before expanding greatly. There is some logic to this argument, but limiting the areas that tourists can backpack in an atmosphere of expanding trekking tourism is a sure way to exceed the carrying capacity of a region. According to Zurick, 'In Nepal . . . jaded trekkers commonly avoid the favourite trekking circuits because they feel that the villages en route have, themselves, become jaded toward the tourist' (1995: 157). Expanding trekking tourism to new areas may be one way to spread the benefits and relieve current pressures on carrying capacities.

Alternative energy sources

Since the ACAP has been opened to tourists, deforestation has become a problem. Because of this problem, energy programmes are a major concern of the ACAP, which has introduced several low-technology methods of conserving energy. Of these, wood-burning stoves, back-boiler water heaters and solar water heaters are the most successful and need to be expanded as much as possible. However, fires are still burned to heat water for tourist showers, and many of the solar collectors are being used at elevations where the water freezes and thus are often broken. Micro-hydroelectricity projects and low-wattage cookers have also been introduced. Hydroelectricity may be the future energy source for Nepal, although it will be expensive to develop and will take some time to alleviate the pressures on the forest. Not only does the electricity need to be generated, but the population, which is now cooking over open fires, needs to be equipped to use electricity. Until such a transition takes place, the forest needs to be protected. The ACAP has sponsored tree nurseries and reforestation projects, which are good but are not keeping pace with fuelwood harvesting.

Kerosene may be the answer for the short to mid-term. It is clear that poor local people are being harmed by the continuing use of fuelwood in lodges. It is also true that there is resistance to the use of kerosene on the part of the national government and the local lodge owners. The loss of foreign exchange to purchase kerosene in a country short of hard currency is troubling for the government, while lodge owners indicated resistance to the use of kerosene because they feel that kerosene is too loud and smelly, and it gets in the food. While these reasons may be valid, they are easily overcome with training and removing cooking to well-ventilated areas. When pressed further, many admitted that it was cheaper to use fuelwood. This is often because lodge owners use family members or employ others to cut wood. Another major problem with the use of kerosene is that it lacks the heat provided by the wood stoves. Porters typically use the wood stove to both cook and stay warm at no cost. Because of the higher cost of kerosene, lodge owners are not likely to let porters or guides cook for free, so tourists are steered away from a lodge that uses kerosene. In a highly competitive environment, this is a real concern for lodge owners.

The ACAP has shown that kerosene can be used successfully to supply all the needs of the trekking tourist. The Annapurna Sanctuary is a section of the ACAP that has a mandatory use of kerosene policy. All cooking and heating for tourists is supplied by kerosene. One lodge charged 50 US cents per day for supplying a kerosene heater under the table on a cold night, but porters were not charged and tourists gladly paid.

The ACAP has sponsored much of the trail repair and bridge construction in the ACAP area to keep tourist routes open. However, the ACAP does not address much of the environmental degradation taking place. In Thorong Phedi (above the tree line), a large lodge has been constructed in a fragile environment where no indigenous population previously lived. In this region of ACAP management, juniper bushes, which are the only vegetation available, are routinely cut to the ground to supply fuelwood for trekkers. In Jharkot, which has a desert-like environment due to its rain shadow location, I witnessed an old woman digging the roots of juniper bushes for fuel. It is a matter of survival for her. In the fall of 1994, I stopped for tea in a wooded area far from any village between Pisang and Manang. The next spring the same spot was devoid of vegetation. The area had been cut back haphazardly and a fire had destroyed much of the rest. Not all people are able to place environmental concerns before survival and some will not put environmental concerns before profit. Community-based development is necessary but areas between villages also need guardianship. There are many examples such as these along the popular trekking trails in the ACAP.

This study has found that while foreign tourists are not without blame, they are more environmentally sensitive than the local population in the study area. The ACAP's minimum impact code should be applauded. However, these suggestions fall short of giving tourists a real option of action. The code rightly advises staying in lodges that use appropriate waste control and fuel management. The problem is that none of the region's lodges use proper environmental controls. It does little good to advise the use of lodges that use kerosene when in most villages none do. What may be needed is an environmental friendliness rating system for each lodge. This information then needs to be distributed to trekkers. Lodges could be inspected periodically by the ACAP and given a rating that must be displayed for tourists. Tourists would need to have an input into the monitoring of this system, because such a posting system could quickly become corrupted in Nepal. This simple scheme would enable tourists to be environmentally sensitive while at the same time it would give lodges economic reasons to adapt environmentally sensitive practices to attract tourists.

Conclusions

Tourists are interested in Nepal, or else they would not travel across the world to trek there. Therefore, well-placed visitor information centres in both Kathmandu and Pokhara would probably be well received. A centre in close proximity to the immigration office in each city, where trekking permits are obtained, would be ideal. Most independent trekkers spend hours queuing to obtain trekking permits at one of these two places. A 15- or 20-minute video educating tourists on environmental problems and cultural practices in Nepal would benefit the environment, society in Nepal, and the trekkers. Tourism is contributing to deforestation problems in the ACAP, and this places a burden on most of the local population. Both group and independent trekking degrade local natural environments by using fuelwood. Food for group trekkers is usually prepared with kerosene. However, it is typical that three support staff accompany each group trekker, and these staff consume fuelwood. Even at the lower consumption rate by the Nepalese, at the three to one ratio each group trekker is responsible for higher fuelwood consumption than each independent trekker. Both group and independent trekking tourism place unsustainable pressure on village forest resources. Fuelwood overuse is a major environmental problem in Nepal, and tourism is adding to this problem. Solutions to these problems can be seen in trekker profiles. Trekkers are mostly from wealthy Western nations, are first-time trekkers, and are willing to protect the environment. Over 80 percent of all trekkers surveyed were willing to pay more for their trek to protect the environment, and over 85 percent of all trekkers surveyed were aware that trekker consumption of fuelwood was causing environmental problems in Nepal. The ACAP should use trekker profiles to facilitate the creation of its tourism policy. Because independent trekking benefits local economies more than group trekking, it should be promoted. Nepal, on the other hand, can use trekker profiles to promote group trekking in regions that do not have the infrastructure necessary for independent trekking.

The solution to the negative impacts of tourists on the forest is to remove tourists from the problem. Hydroelectricity has great potential in Nepal but will take time and capital to develop. Until this happens, the ACAP should pursue a policy that will require tourists and their support staff to use kerosene or other imported fuels. Lessons learned in the ACAP can and should be applied in other regions of Nepal. A policy of fuel self-sufficiency for tourist and support staff may prevent ecological carrying capacities being exceeded. This type of policy has proved to be effective in the Annapurna Sanctuary.

The economic benefits of trekking tourism in Nepal are great. The national government of Nepal benefits greatly from trekker dollars. The regional centres of Kathmandu and Pokhara also do well from trekking dollars, as do the trekking regions of Annapurna and Everest. The villages along trekking routes benefit from trekking dollars. However, the rural poor in trekking corridors are disadvantaged by the consumption of natural resources. Regions of Nepal without trekking routes outnumber the regions that benefit from trekking tourism. The data that this study provides should be helpful to the ACAP and the government of Nepal in planning for the spread of a more ideal form of ecotourism in Nepal, a tourism that benefits local populations and visitors alike without harming the environment in the long term.

The results of this study are important not only to the local populations of Tatopani, the ACAP and the nation of Nepal but also to the entire international community. Lessons learned in the ACAP's search for sustainable development may be applied to other regions of the world, particularly after a thorough evaluation of the pros and cons of the ACAP's experiences. An evaluation of ecotourism gives a clearer picture of the advantages and disadvantages to both the economy and the environment. It is important to refine policies and practices affecting human environmental relations while pursuing economic development.

Chapter 14

Sustainable tourism development and planning in New Zealand: local government responses

Stephen Page and Kaye Thorn

Tourism is not a core element in the planning process despite its apparent economic significance for many localities (Dredge and Moore 1992). While much existing research alludes to tourism as an activity that is planned and may be the focus of planning in many contexts (Inskeep 1991), the reality is that it is not a discrete activity given prominence within the public planning frameworks existing in many countries. Yet it is widely acknowledged that planning and management functions within public sector organisations are the main vehicles for influencing, directing, organising and managing tourism as a human activity with various effects and impacts. Thus the effectiveness of planning for tourism is likely to depend on the extent to which appropriate planning and management functions exist to guide and monitor its development and effects (Heeley 1981).

The growing interest in sustainability as an approach to planning has resulted in a renewed focus on the nature and outcomes of the tourism-planning process for tourism destinations. This renewed interest in the nature of planning for tourism activities and development has also generated a debate on the extent to which tourism is being integrated into the planning process (Dredge and Moore 1992; Pearce 1995). Within a New Zealand context, tourism is a key element of the planning process among regional and local planning authorities, given the government's intention to expand international arrivals from 1.2 million in 1993/94 to 3 million by 2000 (Hall et al. 1997). Expanding international and domestic tourism will place new pressures and demands on the New Zealand planning system, and its ability to recognise and respond to the new development scenarios associated with tourism and tourist activities in different localities has hitherto been neglected by researchers and the public sector. In 1991, New Zealand enacted the Resource Management Act, one of the world's first pieces of legislation that explicitly sought to enshrine the concept of sustainability in planning law. This chapter discusses how the Act has affected the tourism industry and how it has been interpreted and implemented by local and regional government agencies. This chapter illustrates the difficulties that exist with implementation with reference to a survey of local government bodies and regional tourism organisations, and concludes that despite the initial enthusiasm surrounding the Act, substantial areas of uncertainty exist with respect to its longer-term acceptability, the resource consent process, and its application to tourism.

Planning for tourism in New Zealand: the effect of demand factors

The development of New Zealand as a domestic and international tourist destination is well-documented in the academic literature (Pearce 1989, 1990, 1992; Hall *et al.* 1997). The economic contribution of tourism to the national economy was estimated at NZ$8.2 billion for the year ended 1993, half of which comprised expenditure from domestic travel. Tourism is also perceived as offering substantial employment opportunities to many communities at a time of economic restructuring (see Chapter 5). Table 14.1 illustrates the government's targets for tourism growth for the year 2000. Although the feasibility of achieving these growth targets has been questioned and dropped in favour of increased income and expenditure targets (Hall *et al.* 1997; Hall 1998), it illustrates the political emphasis on tourism growth due to its perceived

Table 14.1   New Zealand visitor arrival targets ('000)

Market	1994 (actual)	1995	2000
Australia	386	510	776
Japan	148	215	538
USA	158	227	402
Canada	30	34	49
UK	116	128	182
Nordic	16	25	48
Germany	59	91	180
Switzerland	15	26	52
Netherlands	12	14	27
Taiwan	58	54	153
Korea	62	51	114
Hong Kong	28	34	80
Singapore	24	31	46
Malaysia	16	19	51
Thailand	25	16	38
Indonesia	12	11	27
Other Asia	15	16	38
Other Europe	30	33	64
Other	112	104	68
Total	1,322	1,638	2,934

Source: New Zealand Tourism Board 1995 in Hall 1998.

employment-generating potential now that it exceeds other sectors of the economy as the main source of foreign exchange. When these projections are added to the existing patterns of tourist activity in New Zealand, a number of key features emerge:

- much of the growth in Asian and North American markets to date has been spatially concentrated in metropolitan gateways and a limited number of resort areas such as Rotorua and Queenstown;
- regions other than the gateways and major resort areas are dependent upon the more diffuse travel and spending patterns of Australian and British visitors, which is often VFR (visiting friends and relatives) in character; and
- new niche market segments (e.g. backpacking) are pursuing a more dispersed pattern of tourist activities, though many of these are dependent upon the location of adventure tourism activities.

Research commissioned by the former New Zealand Tourism and Publicity Department (1988) examined the impact of tourism demand on regions, communities and infrastructure. The report generated three scenarios (a base scenario, a concentration scenario and a dispersal scenario), highlighting the likely effects of tourism growth and the implications for planning and management (see Page and Piotrowski 1990 for further details). In fact, Pearce (1990: 40) rightly acknowledged

> how much choice actually is there in view of the impact on demand . . . which lies beyond the influence of the marketers? Given that there is some scope for influencing the market, incorporation of a greater regional dimension would represent a significant change in a long-established practice of fostering the leading growth segments to increase total arrivals.

Pearce's main argument is that tourism marketing in New Zealand has traditionally overlooked the regional spread of arrivals, emphasising total arrivals, seasonality and the targeting of particular market segments. The Tourism 2000 conference held in 1989 saw the task force report that regional tourism development should be market-led and should be the responsibility of the regions (cited in Pearce 1990: 40). The implications are clear: tourism policy by the national government has been reorganised from its pre-1984 public sector role, where investment, marketing, research and development were actively promoted by the New Zealand Tourism and Publicity Department and the New Zealand Tourism Department. The reorganised New Zealand Tourism Board (NZTB) has pursued a market-led philosophy, encouraging the public sector to forge partnerships with the private sector and a greatly diminished role for the central state. There has also been a progressive downgrading of the power and the roles of the government tourism department, which took over the responsibilities of policy advice and issues of tourism in terms of the public good. The Ministry of Tourism was formed in 1991, with eleven staff initially with expertise in planning, resource management and policy advice. By 1995, the Tourism Policy Group was based within another government department, with only seven staff and the resources to undertake very limited strategic planning tasks. This has left a dilemma for the local state: regional tourism organisations are now to determine the future of tourism for their areas in terms of strategic planning without the resources or financial incentives to encourage the development of supply features (e.g. accommodation, attractions and infrastructure). Likewise, local authorities are required to plan for tourism development and activities without additional resources. Therefore, the 1990s are marked by a political climate that is predisposed to rapid growth in tourism arrivals at the national level. But the planning and management implications are delegated to the regional and local planning bodies without any clear national plan to manage and direct the growth to areas able to cope with further development and expansion.

The planning system in New Zealand

In New Zealand, the Resource Management Act (RMA) is the primary law for land use planning and is the main legislation for planning in relation to tourism. It is also the legislation for water and soil management, pollution, waste disposal, coastal management and the subdivision of land for development. The RMA therefore comprises an integrated piece of legislation for the management of the country's environment. It is a relatively recent piece of legislation, being enacted in October 1991. It replaced and repealed 70 other laws, including the previous planning law, the Town and Country Planning Act 1977.

The RMA has one central purpose, which is 'the promotion of sustainable management of natural and physical resources' (RMA 1991: 21), with the term 'sustainable management' defined as 'managing the use, development, and protection of natural and physical resources in a way, or at a rate, which enables people and communities to provide for their social, economic and cultural well-being and for their health and safety' (RMA 1991: 21). Therefore, for tourism, the RMA should encourage public sector planners to adopt a more holistic view of development and the way in which tourism affects the environment and population within a sustainable framework. In this context, sustainable management of tourism involves three main goals:

1 to balance our needs with those of the environment by ensuring that the use of resources does not endanger or irreparably damage any ecological system, including our own;
2 to ensure that acceptably high standards of environmental quality are maintained; and
3 to ensure that the environment and its resources are used in such a way as to protect the ability of future generations to meet their needs.
(Ministry of Tourism 1993: 7)

In a tourism context, the central concept of the RMA is very different from previous legislation in that it emphasises a holistic approach to the determination of impacts. This is supported by the Ministry of Tourism (*ibid.*), which points out that the focus of the Act is on managing the environmental effects of activities rather than the activities themselves. In other words, the RMA does not regulate against certain activities such as farming, manufacturing or tourism; rather, it focuses on the detrimental effects of those activities. Provided that the adverse effects of the activity can be avoided, remedied or mitigated, then it is permitted. The RMA also requires an assessment of environmental effects of any proposed development, including the identification of any potential future impacts (RMA 1991).

The implications of this Act for tourism are significant and wide-ranging, even though the term 'tourism' is not mentioned in the Act. It has been argued that New Zealand's tourism industry depends on the maintenance of environmental quality as a major drawing card for international visitors (NZTB 1991; Hall *et al.* 1997). The clean and green image of New Zealand, based on its large network of national parks and scenic areas currently promoted by the NZTB, is somewhat fragile and can be safeguarded only by managing the environment on a long-term, sustainable basis. The Act is, therefore, directly relevant to both the development and promotion of tourism because it not only has an explicit commitment to the sustainability of the country as a destination but is also concerned with the way in which impacts are managed.

Components of the tourism industry have expressed concern about the RMA as it affects tourism. These concerns are that the time and costs associated with obtaining resource consents under the RMA are proving to be greater than under the old system. It has also been claimed that 'some consent authorities exhibit questionable ethics and decision making skills' (NZTB 1994: 9). Further, there are suggestions that investors, particularly from overseas, are reluctant to invest in tourism in New Zealand, where the time frame, costs and outcomes are effectively outside their control. At a time when there is a need to expand the infrastructure to cope with increased visitor demand, any delay obviously affects the cost of the project. While there is little doubt that the RMA has resulted in delays in some developments, these concerns need to be investigated carefully if the principle of the Act and the need to ensure that quality of the environment is to be maintained.

As the primary planning law, the RMA sets out responsibilities for the central, regional and territorial local governments. The central government provides a national overview and monitoring role with some areas of direct responsibility, particularly relating to coastal management. There is no requirement for the central government to prepare a plan for tourism in New Zealand, although it certainly has the ability to do so. Regional councils have a pivotal role in administering the Act, overviewing the issues in each region and providing the overall framework for all resource management policies within each region (Ministry of Tourism 1993; Dymond 1996; Hall *et al.* 1997).

Territorial local authorities (district or city councils) have prime responsibility for land use management at a local level. These organisations are required to develop a district plan that sets out the objectives of the community for a ten-year period, and the rules for achieving these objectives. It is the local council that is usually the first point of approach for a tourism developer. Local councils are also responsible for providing and maintaining the infrastructure of an area. Although there is a planning hierarchy from central to regional to local government, the RMA essentially delegates the management of the natural and physical resources to the regional and local councils.

The central government and the regional and local councils have responsibilities for tourism through a range of other legislation. For example, a 1990 amendment to the Local Government Act states that a regional council 'may find and co-ordinate the promotion of tourism within the region' (Local Government Act 1990, Sec. 593). Under the same Act, local councils may establish information bureaux and public relations offices, and be directly involved in tourism provided that it promotes community welfare. In addition, the New Zealand Tourism Board Act (1991) identifies international tourism marketing of the country as one of the organistion's main tasks.

It is important to note at this point that around one-third of New Zealand's land mass is held in stewardship and managed by the Department of Conservation (DoC) (Hall *et al.* 1997). This includes many of the country's major visitor attractions such as Milford Sound and Mount Cook National Park (DoC 1994). The DoC is required, under the Conservation Act 1987, to prepare management plans for all conservation land. Further, it has prepared a draft visitor strategy to examine the management of visitors on conservation land (DoC 1994). The local and regional councils are not, therefore, the only agencies responsible for planning for tourism. The integration of practices and information between councils, the DoC and central government would be another interesting area of research to explore in the context of sustainable tourism management (see Hall *et al.* 1997).

There is little doubt that tourism is intrinsically linked into the general planning processes outlined in the RMA, even though it is not specifically mentioned. In principle, the regional council has an overview role with respect to the natural and physical resources as tourism affects those resources. Regional councils may also choose to be involved with the funding and promotion of tourism in a region as outlined above. The local council's involvement is more direct, being the initial contact for many tourism proposals, and setting the rules for the development of the area. It may also be directly involved in the promotion or operation of tourism activities. Therefore, it is evident that a clear range of responsibilities and tasks exist for planning agencies in New Zealand in relation to the administration and implementation of the RMA.

Public sector agencies' responses to tourism and the RMA ———

Within a New Zealand context, there is a limited amount of research on the planning process and its implications in relation to tourism. Studies by Pearce (1985), Davies (1987) and Hart (1992) focused on the situation prior to the RMA (Perkins *et al.* 1993). More recent research by Hall *et al.* (1997) has focused on some of the policy and mangement dimensions of the RMA and tourism planning, while Dymond (1996) undertook a useful survey of the use of sustainable tourism indicators by councils. However, with the above exceptions, little systematic research has been undertaken to assesses regional and local planning organisations' involvement with tourism since the RMA was introduced.

In May 1995, a postal questionnaire was sent to every planning manager in the local and regional authorities within New Zealand. The questionnaire was a lengthy self-completion exercise containing 21 specific questions with multiple responses, which generated over 50 categories of data. It was estimated from the piloting and scoping exercise undertaken to develop the question-naire that it would require approximately 20 to 30 minutes to complete. From a total of 81 questionnaires, 49 replies were received, representing 52 councils (64 percent response rate). Tourism Wairarapa on North Island responded to the survey on behalf of three local councils, and the Western Bay of Plenty Visitor Promotions Organisation responded for two councils, which explains why the 49 replies actually cover 52 areas. Despite these limitations, the data generated provide the first insight into the planning profession's management of tourism via the RMA.

Survey results and discussion

Unfortunately, it is not possible to calculate precise visitor:host ratios for each area that responded, since one cannot gauge the precise volume of domestic visitors as the last New Zealand Domestic Tourism Study was conducted in 1990, there is also relatively little knowledge of the extent of the day-tripper market for some destinations. This remains a substantial weakness in the data available to decision makers and planners who seek to understand domestic tourism. Instead, there is a continued reliance on statistics that measure inter-national arrivals in New Zealand due to their higher economic contribution *per capita*.

Almost all councils responding to the survey considered tourism to be important within their economy. The one exception was from a central North Island locality, where forestry is the dominant industry. Other comments ranged from 'tourism is becoming increasingly important' through to comments that 'it represents 90 percent of all activity within the area'. While there is such a positive response to the role of tourism in the economy, this did not always equate with adequate knowledge of existing patterns, volumes and types of vis-itors and tourists. Only 37 percent of respondents could provide an estimate of the existing volume of both international and domestic visitors into their area and this was mainly for those regions with a population in the 50,000–100,000 category. For the larger areas with a resident population in excess of 100,000 (which included individual localities and districts), the respondents could not identify the patterns of tourism. A more worrying feature is that in the smaller communities of between 20,000 and 50,000, the volume of tour-ism could not easily be identified. Only in those communities of less than 10,000 residents were the tourism volumes easily identified by respondents. In many instances, the figures, particularly for domestic tourism, were dated, reflecting the reliance on the now discontinued Domestic Travel Study.

Only a limited number of respondents (12 percent) could provide an estimate of the likely tourism forecasts for the volume of visitors in their area for the year 2000. Several councils indicated that they expected visitor numbers to double, but they showed no anticipation of considering the effects that these increases may have. The effects of the Sydney 2000 Olympics and the America's Cup defence in 1999/2000 were also considered to be very positive or positive by over 50 percent of respondents. This was particularly so for councils on the Blue Ribbon route in New Zealand (Auckland–Rotorua–Christchurch–Queenstown), although spin-off effects were expected throughout the country. However, such optimism has to be tempered by the fact that of the 26 respondents who felt that the impact was going to be beneficial, only eight had any knowledge of forecasts for their area. This meant that their feelings were not based on a sound understanding of tourism and the impact of hallmark events (Hall 1992b), but on supposition and optimism.

Although there did appear to be a general lack of knowledge of existing and future tourism levels, it was encouraging that over half of all councils undertook either periodic or regular research on tourism. Over two-thirds of those respondents who undertook research were located in areas with resident population of 21,000–50,000 or 51,000–100,000, indicating that the larger and smaller areas found it too expensive or difficult to justify such expenditure on research. In most instances, this research focused on accommodation occupancies, which were used to gauge visitor numbers. This was generally organised through the regional tourism organisation, which in turn is often partially funded by the local councils. It is also acknowledged, however, that accommodation statistics can provide only a broad approximation of the actual volume of tourist visits, particularly in urban areas (Page 1995).

Few councils collected information on visitor activities undertaken in their area and on levels of visitor satisfaction, though council surveys of residents' satisfaction with service delivery are now commonplace in New Zealand. Certainly, variety of research was greater in those councils that are more dependent on tourism. In New Zealand, 63 percent of responding councils did have a specific policy on tourism within their area. However, only 35 percent of councils had developed this into a tourism strategy, although there was an indication that more strategies would be forthcoming. Of this 35 percent, nearly two-thirds of them had developed a visitor management strategy. But knowledge of the numbers and effects of international visitors in specific localities does not appear to have had any bearing on the decision to develop a strategy. The responses therefore showed a considerable disparity in terms of recognising the importance of tourism, and yet councils had restricted information on tourism in the region and a seemingly limited level of commitment to the formation of a tourism policy or strategy. The obvious exceptions to this were those councils whose areas are almost entirely dependent on tourism – their approach appeared more cohesive and thorough.

While the questionnaires were sent to the planning department of each council and the majority of responses did come from these departments, a significant number of questionnaires had been forwarded to the promotion or marketing divisions of the council. It is difficult to determine exactly how the responsibilities for tourism were allocated within each council, but this does suggest that tourism is being identified as an existing activity requiring promotion rather than as something that requires strategic planning. The majority of responses came from planners/resource managers (69 percent) followed by tourism marketing and promotions officers (12 percent), economic/tourism development officers (8 percent) with the chief executive officer replying in 4 percent of cases and 7 percent of respondents not specifying their job title. This indicates that a range of locations did not consider or have the resources to appoint a specific person with responsibility for tourism issues. This is compounded by the fact that those planners dealing with tourism will normally have no formal qualifications in tourism, as tourism is treated as just one other form of economic development. This is an important underlying factor reflected in the responses received and the degree of precision and understanding respondents show in relation to tourism. It may also explain why planners are not necessarily aware of the major data sources available to gauge the impact of tourism in their locality.

A range of councils did recognise that there are capacity issues to be addressed within their area, either currently or anticipated. These issues focused on lack of accommodation (41 percent), lack of adequate infrastructure (29 percent) and the importance of sustaining the environment (31 percent). With a few exceptions, however, there appeared little consideration of how these capacity issues could be managed. Limiting visitor numbers was suggested by several councils, although they appeared to have no visitor management strategies. One council, which has specific problems in this respect, was considering commercial rating differentials, infrastructure payments and tighter design controls as strategies to limit growth.

This recognition of the need to minimise the impacts of tourism was also identified as a requirement of the RMA by 41 percent of respondents. A further 20 percent recognised the need to ensure the sustainable use of the physical and natural resources, with many of these resources being important to tourism. Not all councils recognised this connection, so not all respondents saw the natural corollary that tourism needs to be planned. Some councils admitted this was an issue that they were currently addressing, while others considered that the RMA had no effect on the need to plan for tourism. One council went as far as to say that it had yet to identify a tourism activity that required a planning response. Other councils identified some of the negative aspects of the RMA legislation, including 'hindering development' and 'costing the ratepayer an enormous sum of money and causing delays'. Interestingly, less than 50 percent of those areas progressing some form of development had a formalised tourism strategy, though those areas with development proposals were more likely to be able to refer to forecasts for tourism.

Councils drawing upon published data reported a heavy reliance on government publications as the source of information used in the preparation of their district or regional plans. Public submissions, other council reports and local knowledge were also important. While these sources are clearly relevant, there did seem to be a focus on local information as opposed to a national or global picture. This has implications for an integrated approach to planning, and more specifically, tourism planning, indicating that many councils adopt a rather insular and inward-looking approach to tourism, given the absence of a national approach to provide guidance on tourism in New Zealand.

Public consultation is a requirement of the RMA and a vital element in the democratic process for the formulation of policy in the public sector. Most councils reported that they had an extensive programme of public meetings, key group consultations and submissions. There appeared to be little participation by the tourism industry in this process. While there is no suggestion that councils should specifically target the tourism industry, the opportunities for involvement in the planning process are explicit and available to all interested parties.

Implications for sustainable tourism planning

From the survey results, it is apparent that as tourism assumes increased economic and social importance in New Zealand, the local and regional economies are left to face the realities of future tourism development scenarios. Most local authorities and regional councils have not actively taken stock of the situation in their locality or pursued a focused direction for tourism, because of the absence of any guiding principles from national government or because they do not perceive a need that justifies spending their financial resources. The RMA may provide a perceived framework for the sustainable management of natural and physical resources, but in a tourism context it has not encouraged more than a narrow notion of locality and impacts that are not sufficiently integrated into the wider planning process. As a planning tool, the RMA has failed to establish whether tourism should actually be expanded, constrained or deterred in specific areas. With the RMA focus on individual project impacts, many local planning bodies have adopted a project-by-project approach to tourism in the absence of an integrated planning framework for tourism informed by policy and strategy formulation and development. This is particularly worrying as areas seize perceived opportunities afforded by tourism without a clear notion of future visitor impacts (see Hall et al. 1997). In this context, Dredge and Moore's comments (1992: 20) are significant to the New Zealand situation:

> As we move into the 1990s, increased tourism growth will result in greater challenges for the integration of tourism and town planning. These challenges will be brought about by the need for the development of attractions, transport, support services and infrastructure to cater for increased visitor numbers, and the implications this will have for land use planning. . . . Planners have a responsibility to meet the challenges offered by the growth in tourism and to understand how their activities affect tourism.

The outcome of this research certainly reinforces these conclusions, highlighting both the dearth of knowledge of the scale, volume and economic significance of tourism, and the lack of involvement most local and regional governments have in understanding the real implications of basing future economic prosperity on tourism. Dredge and Moore's (1992) framework for the integration of tourism into planning on a state level is certainly relevant to the local and regional councils throughout New Zealand.

This study however, is perhaps more disturbing because of the implications its findings have at the national level. Only the local and regional councils have any role at all in terms of managing tourism growth in New Zealand, and it is apparent that even there, involvement is limited. The central government continues to market New Zealand internationally through the NZTB. However, the only agency responsible for assessing the situation once the visitors are in the country is the Tourism Policy Group, which lacks the resources to execute this task. In other words, there is no national focus for dispersing visitors, there is no indication of how these visitors are going to come or how they will travel when they arrive, a point which reinforces Pearce's (1990) conclusions on the absence of any national framework for spreading the benefits of international and domestic tourism to a wider range of localities. There is limited knowledge of the effects of tourism, little acknowledgement of the incremental effects of tourism and a political attitude that is predisposed to growth without public sector investment.

The research findings also indicate that there is a paucity of disseminated research findings for both decision makers and planners in the public sector. Furthermore, the lack of national commitment towards tourism research following the restructuring of central public sector responsibilities for tourism highlights the philosophy of tourism development with limited state investment in planning for the effects and impacts. The Tourism Policy Group (1995, 1996) acknowledges the efforts of a limited number of academics, the contribution of commissioned research and a wide range of student research projects in tourism research output in New Zealand. However, the existing state of knowledge on international and domestic tourism in New Zealand is no longer keeping pace with that in other competing destinations. The implication here is clear: with inadequate research data and material upon which to base decisions, public sector planning may not necessarily be able to understand fully the effects of actions under the RMA, since research is a vital part of the planning process for tourism.

Conclusions

The primary aim of this study was to examine the involvement of local and regional councils in tourism planning in their areas and the relationship with sustainable tourism management. It has clearly identified the need for both more information and a greater focus on integrated planning, thereby identifying a glaring gap in the management of New Zealand's tourism industry, development and visitor activities. It is acknowledged that it costs money to plan effectively, even though, in the current environment of public sector financial stringencies, additional costs are not encouraged. In fact, some proponents naively argue that tourism is still growing – so why spend money on a problem that does not exist? On the contrary, until the New Zealand government prepares an integrated master plan or guidelines on the growth of tourism into the twenty-first century, the outlook is bleak. Implementing a future based on the principles of sustainable tourism will be little more than political posturing in the absence of a vision for the future growth and management of tourism in New Zealand.

Sustainability demands an understanding of the effects of tourism, and the need for leadership, resources and coordination. Without central government commitment to these issues, the principles of sustainability embodied in the RMA and their relationship to tourism will remain rhetoric rather than reality. It is highly likely that the central government will place the onus on the local and regional councils as the final arbiters of sustainable tourism. In the absence of central government commitment to these principles, through resource allocations to fund initiatives and integrated tourism planning, local and regional governments will have to prioritise scarce funds to address tourism issues. The question here will be: what is the opportunity cost of focusing on tourism in a climate of limited funding? For the central government to admit to the need for a national tourism plan is almost a recognition that existing policy is poorly developed, based on false assumptions and driven by commercial interests.

Tourism planning seems destined to remain a reactive response to problems and pressures generated by tourism in New Zealand, leaving the localities to find their own solutions to tourism's impact. Meanwhile, the central government continues to fund the promotion and marketing of New Zealand overseas to maintain the intended growth in international arrivals. This is certainly not a clear example of a commitment to sustainable tourism planning by any measure: it is a short-term view that is motivated by the economic gains that visitor expenditure generates for the national economy to assist in the balance of payments. Without some degree of investment in the future sustainability of tourism in New Zealand's regions and local areas, guided by a national vision and geographical policy towards development and growth, the future image of a clean, green and unspoilt tourist destination is likely to be eroded in the next century.

Acknowledgements

An earlier version of this paper was first presented at the IGU Regional Conference on the Geography of Sustainable Tourism in Canberra in September 1995. The authors would like to acknowledge the financial assistance provided towards this research by Massey University.

Chapter 15

Sustainable urban tourist attractions: the case of Fort Edmonton Park

Thomas D. Hinch

'It is somewhat of a paradox that although urban areas have been recognized as one of the most important type[s] of tourist destination[s]' (Law 1993: 1), they have, to a large extent, been excluded from discussions about sustainable tourism. This omission of urban tourism is difficult to explain, but it is at least partially due to the dynamic and complex nature of the city and the often fragmented nature of its tourism function. Whereas unsustainable tourism development in pristine natural environments often results in dramatic failures, thereby drawing the attention of researchers and planners, the consequences of similar practices in an urban setting frequently go unnoticed. Despite this tendency, the results of unsustainable tourism practices in an urban setting are similar to those in a natural setting: tourism resources deteriorate, visitation wanes, community support for tourism decreases, and the industry becomes unprofitable.

Sustainable tourism development in the city is, therefore, just as important to understand as it is in other geographic areas. Given their central place within urban destinations, attractions would appear to play an important role in the pursuit of sustainable urban tourism. This chapter explores these themes by, first, reviewing underlying concepts of sustainable tourism in an urban setting; second, articulating the role of managed tourist attractions in fulfilling the objectives of sustainability; and third, using a case study of Fort Edmonton Park (FEP) in Alberta, Canada, as an illustration of these concepts and roles. The case study data sources include existing studies and records as well as in-depth interviews with selected managers and staff.

Sustainable urban tourism

There has been widespread support for the concept of sustainable development as 'development that meets the needs of the present without compromising the ability of future generations to meet their own needs' (World Commission on Environment and Development 1986: 43). Sustainable tourism is consistent with this broader concept (e.g. Bramwell and Lane 1993; Nelson, Butler and Wall 1993; Pearce 1995) and with similar concepts under a variety of labels (Butler 1990). One of the most useful variations of this concept is offered by

Murphy (1985) in his ecological model for community-based tourism planning. This model presents four elements that are fundamental to sustainability, including:

1 the protection of the destination's resource attractions;
2 positive resident assessment of tourism;
3 visitor satisfaction with their experience; and
4 an acceptable return on investment for operators within the tourism industry.

If these conditions are met, tourism will be sustainable in the long term and consistent with the broader concept of sustainable development.

It is only relatively recently that tourism researchers have begun to study tourism in urban areas (Ashworth 1989, 1992; Ashworth and Tunbridge 1990; Page 1995). Cities have generally been thought of as areas of origin for tourist flows, while non-urban areas have been considered the areas of destination (Stansfield 1964). Christaller's (1963/64) assertion that urban residents tend to seek their recreation in non-urban peripheral areas would seem to have been shared by most tourism researchers until the mid-1980s. However, if tourism is viewed in a broader context than simply holiday travel, Christaller's (1966) own theory on central places suggests that urban areas can be expected to serve as major destinations. People travel to cities for a variety of reasons, including to visit friends and relatives, to access transport networks, to conduct business and attend conferences, for personal reasons such as medical treatment and education, and to participate in a wide variety of cultural, artistic and recreational activities (Blank and Petkovich 1987). Given the concentration of these opportunities in urban areas, Law's (1993) claim that cities are the most important tourist destinations in the world is convincing.

Urban tourism resources are not evenly distributed among cities. In addition to quantitative differences associated with the scale of the city, there are also qualitative differences. Each city has its own blend of characteristics that contribute to its unique sense of place. Garnham (1985) argues that this character or genius loci is responsible for the bond that forms between people (visitors or residents) and a place. A broad range of traits contribute to this sense of place, including but not limited to architectural style, climate, natural setting, spatial relationships, cultural diversity and societal values. To a large extent, it is this sense of place that forms the urban attraction. Successful urban tourism destinations are able to foster a unique and desirable sense of place from the perspective of visitors. Three fundamental dimensions of place in an urban setting are:

1 the built environment;
2 the natural environment; and
3 the cultural environment.

The built environment is epitomised by distinctive architecture, historic buildings and districts, sports and cultural centres, shopping areas, restaurants and entertainment, and even the industrial areas (Jamieson 1990; Inskeep 1991). From the perspective of sustainability, one of the biggest threats to this dimension is the trend towards the homogenisation of urban design. As the built dimension of the urban setting becomes more and more similar within and between cities, Relph's (1976) concept of 'placelessness' becomes more tangible. Built environments that are indistinguishable from the home environment of potential tourists will not be attractive destinations.

While urban areas tend to be dominated by built form, the natural areas found within the city offer a necessary counterbalance. A spectrum of natural areas represents an integral part of an urban destination's appeal (Barke and Newton 1995). Unfortunately, open space within the city is often mistakenly viewed as underused space, and as a result it is constantly under the threat of more intensive development. This threat typically comes in the form of transport development and, somewhat ironically, it often comes in the form of intensive tourism development. Hall (1994b) provides a good illustration of this incremental erosion of urban green space through infrastructure development for the hosting of major events, such as the Olympics.

The final dimension of the urban setting to be singled out is the socio-cultural environment. The cultural patterns, traditions and lifestyles associated with a place can form integral elements of its sense of place and ultimately its attraction. This dimension is also one of the most sensitive in terms of sustainability in the context of tourism due to the challenges of commodification (Greenwood 1989). The danger exists in that through the very process of marketing local culture for the consumption of visitors, the essence of that culture and its contribution to the sense of place may be lost.

Tourist attractions in the urban setting

Like any tourist destination, a successful urban area is characterised by a mix of interrelated components. Jansen-Verbeke (1986) has suggested that the urban tourism product is made up of primary elements (activity places and leisure settings), secondary elements such as accommodation and restaurant services, and additional elements such as parking facilities and tourist information. Sustainable tourism requires the successful integration of all these elements. Each is an integral part of the overall system, but it is the attraction that the other elements are built around.

Urban attractions can be categorised in a similar fashion to Jansen-Verbeke's division of primary elements into activity place and leisure setting. The leisure setting is most closely associated with theoretical definitions of attractions (Lew 1987; Gunn 1988a; MacCannell 1976). This approach is exemplified by Leiper's (1990: 371) definition of a tourist attraction as 'a system comprising three elements: a tourist or human element, a nucleus or central element, and a marker or informative element'. A tourist attraction comes into existence when the three elements are connected. In an urban context, the nucleus of the urban attraction would be the leisure setting and the sense of place as engendered through the built, natural and socio-cultural dimensions of the city; the human element would be the visitors; and the markers would be highlighted by promotional efforts that communicate the desired sense of place that is being marketed.

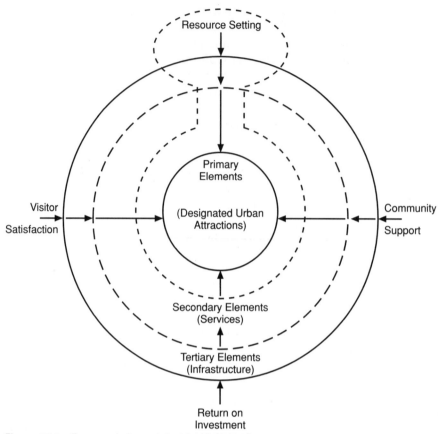

Figure 15.1 Framework for sustainable urban tourism.

Activity places represent specific manifestations of urban tourist attractions. Pearce's (1991: 46) definition of a tourist attraction as 'a named site with a specific human or natural feature which is the focus of visitor and management attention', fits this perspective. Defined in this way, attractions are the points at which urban resources are processed. They represent an acknowledged site of controlled commodification in a complex urban tourism system. In contrast to the attraction of the leisure/resource setting, specified urban attractions may generate a significant level of direct revenue. Because of their explicit role in the tourism system and their manageability, these designated urban attractions serve as useful focal points for the examination of sustainable tourism in the city.

Figure 15.1 illustrates a framework for examining sustainable urban tourism. The core of the framework is the designated urban attraction (DUA) or activity space that is the target of visitor and management attention. The broken line surrounding the designated attraction reflects the roots of the DUA in the broader attraction of the resource setting, the preservation of which is a condition for sustainability. In a general sense, the DUA is a processed urban resource, while the other elements in the resource setting have not been consciously refined for tourism consumption. The next layers of the framework include the balance of the urban tourism product inclusive of secondary (services) and tertiary (infrastructure) elements. Broken line boundaries indicate that the distinctions between these categories are not absolute. Finally, Murphy's (1985) four elements of community-based tourism planning are integrated into the urban tourism framework as conditions for sustainability. This framework will now be used to examine Fort Edmonton Park as a sustainable urban tourist attraction.

Fort Edmonton Park

Fort Edmonton Park (FEP) is a living history interpretive centre in Edmonton, Alberta. Located on a 65-hectare site on the south bank of the North Saskatchewan River in the middle of a major river valley park system (Figure 15.2), FEP is recognised as a major attraction in this provincial capital of approximately 850,000 people. Purported to be Canada's largest urban historical park, this attraction depicts living and working conditions of the fur trade and pioneer periods of Edmonton's history through interactive programming on a site featuring a collection of historic buildings from throughout the city. Fort Edmonton Park is owned and operated by the city of Edmonton and is supported by a combination of site-based revenues, sponsorship and municipal tax dollars. A variety of community-based groups, such as the Rotary Club and the Fort Edmonton Historical Foundation, have been instrumental in FEP's development and operation since its opening in 1974. Its guiding mission is

Figure 15.2   Fort Edmonton Park.

to provide 'diverse opportunities for people to learn, grow, and enjoy themselves through the conservation, animation and experience of Edmonton's history' (Fort Edmonton Park (FEP) 1996a). A replica of the fur trading fort that operated in the Edmonton area in 1846 forms the major element. This fort is intended to serve as the starting point for a typical patron's visit, followed by a 'walk through time' along three streets featuring restored buildings from 1885, 1905 and the 1920s. Fort Edmonton Park also contains land used to support a working period farm and open space that reflects the landscape of the periods being depicted. Historic modes of transport are used throughout the park, including a steam train, a variety of horse-drawn carriages and early motor vehicles. Extensive interpretive programming is employed as a way of animating the site. Other on-site features include food services, customised programmes, special events, merchandising, volunteer programmes, and rentals. In the context of the natural environment, FEP is a key component of the river valley park system that bisects Edmonton. While parts of FEP are developed intensively, it has been designed to be relatively unobtrusive and complementary to the natural landscape. Finally, programming at FEP has emphasised Edmonton's cultural past in a way that is relevant to present users. A visit to the site is meant to be entertaining as well as educational.

Resource integrity

The relocation of a historic building is accompanied by a significant loss of contextual meaning, but in many instances, the alternative is to lose not only the local context but the entire facility. In Edmonton, during the oil-fed construction boom of the 1950s to the 1970s, older structures were routinely being demolished to make way for modern developments. Legislative programmes and government-sponsored incentives resulted in the on-site preservation of several historic buildings in Edmonton, but certainly not the majority. FEP became a depository for representative samples of these structures, which otherwise would not have been preserved. Relocation created a focused urban attraction that facilitated management and visitor attention, which was directed toward the preservation of these historic resources and their use as leisure amenities. FEP was designed to facilitate the commodification of these resources in a sustainable fashion. Key strategies in this regard included the site's physical design and the fundamental pursuit of authenticity.

The spatial configuration of FEP conforms to Gunn's (1988b) description of a tripartite attraction design. At the nucleus of the park are the historic buildings and programme areas as previously described. An inconspicuous but functional security fence surrounds this nucleus, along with a substantial amount of open space within the fenced parameter. This zone corresponds to Gunn's 'inviolate belt'. Controlled access is provided through a railway station next to the parking lot. After paying their admission, visitors usually board a steam train, which takes them on a five-minute trip to the fort. An added measure of protection is provided by the surrounding river valley parkland in which FEP is located. Although infrastructure in the form of parking lots and roadways are provided in association with FEP, intensive development is generally excluded from the river valley. The surrounding residential neighbourhoods are spatially and visually removed from the site. Beyond this buffer, commercial services such as petrol stations can be found, corresponding to Gunn's 'zone of closure'. As pressure has increased to reduce the tax dollar support, commercial activity within FEP has increased, including food services, merchandising and facility rentals.

The authenticity of the buildings and programming is safeguarded through the development of strategic plans for the site (FEP 1984). The park employs a researcher whose primary responsibility is to clarify the historical context for the buildings and programmes. In addition, the Fort Edmonton Historical Foundation functions as an independent check on issues of authenticity. When considering the addition of new historic buildings to the site, a set of specialised criteria is used to assess their suitability (FEP 1996b).

Despite these guidelines and intentions, it is recognised that authenticity is a relative term. Increasingly, FEP is facing budgetary restrictions that make it difficult to maintain existing standards of authenticity. Authentic materials and skilled craftspeople are becoming rarer and more expensive at the same time as the political willingness to fund 'soft services' is decreasing. Over the past few years, some structural maintenance has been deferred due to budgetary restrictions. Such deferrals cannot take place indefinitely and, in the interest of preservation and budget realities, some of these maintenance repairs may not be as authentic as they have been in the past. Other authenticity compromises include the development of selected 'out of time' areas such as modern public washrooms. Such spaces are needed not only to meet current health regulations but also to meet the demands of visitors. Similarly, low-profile ramps have been installed to provide wheelchair access to many of the buildings, which did not have such access when they were constructed. When asked, site interpreters explain the alterations to authenticity along with the rationale for these changes. Accurate portrayal of race relations is another challenge in terms of authenticity. Today's societal attitudes and norms are not consistent with those that existed in earlier periods. Re-enactments of past relationships between native peoples and the settlers of European ancestry tend to focus on what society currently views as acceptable behaviour, and any exceptions are carefully explained.

At the heart of these challenges is the distinction between a living history museum and a theme park. Further incremental erosion of the authenticity at the park could theoretically shift the balance from its present status as a living history museum to a fantasy-based theme park. The temptation to present an entertaining, but inauthentic, version of Edmonton's history in the form of a Hollywood presentation of the 'wild west' has not been acted upon. Management believes that authenticity is the cornerstone of FEP's competitive advantage. Not only is there intrinsic value in preserving this authenticity, but there is economic value in that no competing destination can copy this sense of place. Yet, it is equally true that the entertainment value of a visit to the site is seen as critical to its sustainability. To this end, management has emphasised the living component of FEP by programming a wide range of interpretive situations, including historical vignettes and hands-on activities for visitors.

Visitor satisfaction

Traditionally, Fort Edmonton Park has used attendance as an indicator of visitor satisfaction. During its first year of operations in 1974, FEP hosted 92,737 visitors (FEP 1996a). Since that time, visitation has increased, with a 1996 attendance of 202,817 visitors (Figure 15.3). The distribution between local and out-of-town visitors has been approximately six locals for every four out-of-town visitors. Yet a significant portion of the visitors from Edmonton were in fact hosting friends and relatives during their visit, thereby highlighting the importance of tourism.

In the early 1990s, management initiated annual surveys to measure trends in customer satisfaction and to track changing customer needs. Exit surveys of visitors measured overall satisfaction by asking respondents if they were 'completely satisfied', 'very satisfied', 'fairly satisfied', 'not satisfied' or 'not at all satisfied' with their visit to the park. The portion of respondents who answered completely satisfied or very satisfied was 93 percent in 1992, 100 percent in 1993, 93 percent in 1994, 98 percent in 1995 and 91 percent in 1996 (FEP 1997a). Primary research identified two basic customer profiles. The first makes up about 51 percent of visitors and consists of family groups with an average size of three to four members, annual household income from C$25,000 to C$50,000, and university-educated parents. The second customer profile are adults aged 19 to 44 without children, and they are patronising FEP while in the city to visit friends and relatives.

Indicated satisfaction has remained high, although it has fluctuated from a perfect score of 100 percent in 1993 to a low of 91 percent in 1996. While a variety of factors such as weather and survey methods may explain this fluctuation, the cumulative impact of budgetary restrictions may be beginning to impact the visitor's on-site experience adversely. Management has recognised that reinvestment in FEP is needed to maintain the excellent customer satisfaction levels that have been demonstrated over the past five years. If customer satisfaction were to drop significantly, the repercussions in terms of lower attendance and decreased revenues could have a spiralling effect, making it very difficult to maintain existing standards of quality.

Community support

Fort Edmonton Park has maintained a good relationship with the surrounding neighbourhood. Its initiatives in the area of after-hours facility rental and special event programming did cause some concern in the early 1990s. These initiatives were pursued as a strategy for introducing new revenue streams. A special events tent was even erected to house events such as convention banquets that were too large for existing facilities. While this venue proved popular with clients, complaints of late-night noise began to be lodged by individuals

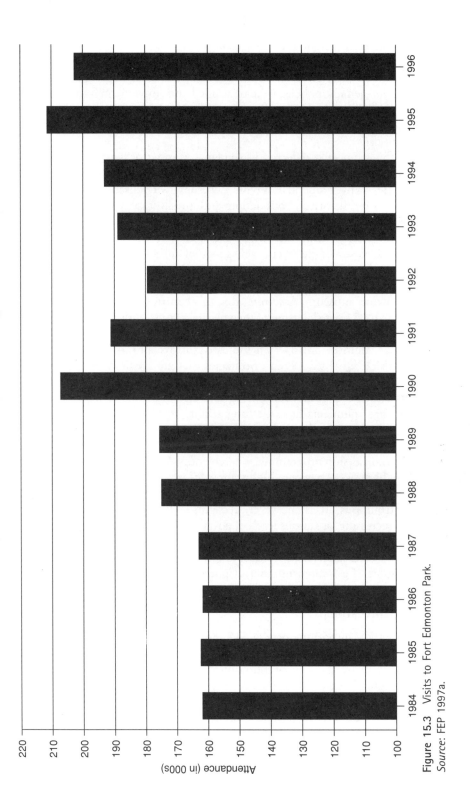

Figure 15.3 Visits to Fort Edmonton Park.
Source: FEP 1997a.

in the surrounding neighbourhoods (*Edmonton Journal* 1991). These complaints were addressed by a monitoring programme in the neighbourhoods, which confirmed the legitimacy of the complaints. Acoustic design modifications and management policies were developed and implemented to solve this problem. There have been no other prevalent complaints over the past five years. Indeed, indicators suggest that FEP now enjoys a high degree of community support.

Part of this support is reflected in the large number of local visitors to the site, but an even stronger indicator is the high level of volunteer involvement. Not only are many individuals involved through organisations such as the Fort Edmonton Historical Foundation and the Rotary Club, but records show that there were at least 849 volunteers, who contributed a total of 34,745 hours of their time directly to the operations in 1996 (FEP 1997b). These volunteers served in a very wide range of roles from costumed staff to behind-the-scenes workers, and their involvement has been absolutely crucial to the functioning of FEP, especially in terms of providing the animation that makes it a 'living' museum. Assuming that this contribution is valued at C$11 per hour, the total financial value of the volunteer contribution was C$382,196 in 1996.

Surveys of volunteer satisfaction indicated an overall satisfaction rating of 87 percent in 1996 (FEP 1996a). Strategies that are credited with obtaining high satisfaction levels include formal recognition of the contribution of volunteers and the ongoing monitoring and attention to volunteer needs. It is also felt that volunteers have a strong affinity to FEP and its fundamental purpose of presenting a 'living history' of Edmonton. Continued support by this group depends on their continued perception that FEP is fulfilling this purpose. A sponsorship programme, whereby outside agencies provide financial support to FEP, is seen as an important way of capitalising on this public goodwill while at the same time decreasing the attraction's dependence on tax dollar contributions. It is, however, recognised that competition for these sponsors is increasing, as various levels of government are reducing their financial support on many fronts. Not only must Fort Edmonton Park compete with other leisure-based agencies but it must also compete with hospitals and schools for the support of a limited number of potential sponsors.

Return on investment

Since the park's inception, over C$15 million has been spent on capital development through the fund-raising efforts of a variety of volunteer groups and expenditure by the city of Edmonton (FEP 1996a). Annual operating costs have regularly exceeded annual revenues (Table 15.1). In 1996, the most recent year for which data are available, the operating expenditures of the park were just over C$2,848,000, while revenue was C$1,254,000. Cost recovery was calculated at 44 percent, which represents a 9 percent increase from 1984, when only 35 percent of expenses were recovered through direct revenue. This growing level of cost recovery is consistent with FEP objectives to strive towards

Table 15.1 Financial performance of Fort Edmonton Park (figures in C$'000)

	1984	1985	1986	1987	1988	1989	1990	1991	1992	1993	1994	1995	1996
Expenses	1,398	1,400	1,782	1,842	1,870	2,114	2,488	2,701	2,638	2,763	2,661	2,780	2,848
Revenue	496	500	663	619	663	754	889	879	831	962	1,022	1,129	1,254
Cost recovery	35%	36.5%	37.5%	34%	35%	36%	36%	33%	32%	35%	38%	41%	44%
Tax support	902	900	1,119	1,223	1,207	1,360	1,599	1,832	1,807	1,801	1,639	1,651	1,594

Source: FEP 1997a.

increased self-sufficiency (Pannell Kerr Forster Management Consultants 1990). It also reflects the concerted efforts of management to increase revenue through a variety of marketing initiatives while at the same time decreasing expenditure. Despite these efforts, it is clear that Fort Edmonton is not currently self-sufficient in financial terms.

Discussion

This review of FEP reveals that three of the four conditions of sustainability are definitely being met. FEP has successfully commodified Edmonton's built and cultural heritage in a manner that retains the authenticity of these resources and allows for their preservation in the long term. The preservation and interpretation of these resources through FEP has contributed to a unique sense of place in Edmonton at a time when these resources have been disappearing elsewhere throughout the city. Visitor satisfaction and community support have also been measured and demonstrate very positive outcomes. Customer satisfaction levels of over 90 percent for each of the last four years and community support as measured through the high levels of volunteer participation on the site are clear indicators that these conditions of sustainability are being achieved.

Despite the successes in these areas, FEP's performance related to the final condition for sustainability is not clear. FEP would not exist in its current form without significant tax-based support to make up the difference between operating revenue and operating expenses. This operating shortfall exists even in the face of aggressive management efforts to reduce this dependency. However, at least two additional questions must be answered before passing judgment on FEP's overall sustainability. The first relates back to the framework of sustainable urban tourism. One of the key aspects of the framework was that the conditions for sustainability need to be considered in the context of the whole urban destination. Traditionally, tourist attractions have not been the primary revenue generators within this system, although there are certainly examples of commercial attractions that generate significant revenues. Yet even in the case of Edmonton's major commercial tourist attraction, West Edmonton Mall, one of the largest malls in the world, the true return on its leisure amenities goes beyond the revenues generated at each amenity to those generated in the surrounding retail stores and services throughout the mall and beyond. Given that FEP functions as a major attraction within a larger tourism system, the broader economic impacts must be considered. For example, how significant is a traditional banquet in the unique facilities of FEP for convention delegates visiting the city when compared with yet another banquet meal in a generic hotel dining hall? What impression of Edmonton does FEP instill and how will it influence future travel behaviour? Completing an economic impact study of this nature is much more complex than simply assessing the financial statements of the attraction at the end of the year, but it provides a more complete assessment of the true return on investment from the destination's perspective.

The second important consideration is that FEP is also a major recreational and cultural amenity for city residents. A value judgment needs to be made about public interest that goes beyond economic considerations. Are there non-economic benefits for the residents of Edmonton associated with this facility that justify the use of tax dollars for the operation of the site? To the extent that the answer to this question is yes, the city's operating contribution should be viewed as the cost of these benefits rather than a subsidy.

Both of these questions highlight the subjectivity associated with determining an acceptable rate of return. In the context of a privately owned commercial venture, an acceptable rate of return would be one in which the owners earn a profit that is commensurate with the level of risk to which they are exposed (Gunn 1988a). Assessing this dimension of sustainability becomes largely an accounting exercise. While the exact level of profit that is acceptable may be somewhat subjective, at a bare minimum the operational revenue must usually be equal to operational costs.

In the case of a public sector attraction, the determination of an acceptable rate of return is much more subjective. This is because the decision must be made from the perspective of the destination community as a whole (Murphy 1985). From the public perspective, the net economic impacts of the entire tourism system need to be considered, along with the even more complex non-economic benefits and costs. Assessing the sustainability of public sector attractions like FEP, therefore, shifts away from the realm of accounting to the realm of politics.

Conclusion

Fort Edmonton Park appears to be clearly sustainable in the context of three out of the four conditions identified in the framework for sustainable urban tourism, although an injection of capital is likely to be needed to maintain high performance levels. Sustainability in the context of the fourth condition, concerning an acceptable return on investment, is not as easy to judge. A political decision must be made that is inclusive of, but goes beyond, an analysis of FEP's annual financial statements.

At the time of writing of this chapter, Edmonton is controlled by a fiscally conservative council facing a reduced tax base. The city administration is currently in the process of organisational restructuring, and it has already merged the Department of Social Services with the Department of Parks and Recreation, which includes FEP. The new department is called Community Services, and further reorganisation is currently being undertaken within this entity. Basic governance questions about major attractions like FEP are being explored, with possible outcomes ranging from the maintenance of the status quo to the unconditional sale of these assets to private investors. Given the current environment, the likely outcome will be a shift along the continuum towards greater public sector involvement. The magnitude of this shift is difficult to predict. Financial considerations of the current tax dollar contribution to the operation of FEP are likely to be a primary consideration. These considerations should, however, be viewed in the context of the discussion of sustainability just presented here. Whatever delivery system is selected, it should be chosen with due consideration to the implications throughout the whole urban tourism system. An equilibrium position should be sought that satisfies all four conditions for sustainable urban tourism. A delivery system for urban tourist attractions that is solely driven by short-term financial considerations may jeopardise existing successes in the area of visitor satisfaction, community support and resource protection. Potential loss of the volunteer base, the erosion of existing levels of customer satisfaction and the potential loss of the competitive advantage associated with unique heritage resources must be carefully assessed and factored into these decisions. Short-term solutions that 'solve' the existing financial imbalance must also maintain or improve on the performance related to the other conditions needed for sustainable urban tourism. If these other conditions are not met, the cost savings of any intervention are likely to be short-lived. Further research on the ability of a variety of service delivery frameworks to meet all four of these conditions is needed.

Acknowledgements

I would like to thank the management and staff of Fort Edmonton Park, who were so generous of their time and information. The arguments and conclusions contained in this chapter are the author's and are not necessarily shared by the management and staff of Fort Edmonton Park.

Chapter 16

The geography of sustainable tourism: lessons and prospects

Alan A. Lew and C. Michael Hall

The concept of sustainable development and the idea that sustainability might inform the goals and methods of tourism development have given new voice to concerns over the urgent environmental issues of our time. As seen through the authors contributing to this volume, the issues that the sustainable tourism debate raises are central to the interests of geography and geographers in that they focus on the character of places, how places are created and changed, and how places are used in a rapidly changing world. In particular, the geographic focus has taken up the issue of the protection of cultural diversity and heritage in the face of economic development and changing economic relationships. This chapter attempts to summarise the state of geographic inquiry into sustainable tourism by reviewing the major findings and lessons presented by the authors in this volume.

Lesson 1: *Sustainable tourism represents a value orientation in which the management of tourism impacts takes precedence over market economics – although tension between the two is ever present.*

In Chapter 1, we noted that while geographers have held a traditional interest in environmental, regional and spatial phenomenon as these relate to tourism, their interests also reflect broader societal concerns in the way that these issues are addressed. Thus, while the concept of 'sustainable tourism' is relatively new in the popular lexicon, the ideas and ideals that it embodies are quite old. In reviewing the historical antecedents of sustainable development, Hall (Chapter 2) noted that from its roots in the nineteenth century, conservation has been defined, and indeed justified, in relation to economic imperatives. And because natural environments are major tourist attractions, from the outset tourism has been a major factor in the conservation debate. Both Hall and Butler (Chapter 3) identified what may be an underlying theme for much discussion on sustainable tourism – that from the conservationist viewpoint, the issue of controlling visitation tends to take precedence over economic considerations. Hull's (Chapter 12) and Pobocik's and Batulla's (Chapter 13) descriptions of environmental tourism were good examples from the case studies section of the book on the importance of conservation values in preserving both the natural environment and the communities in a destination so that a sustainable form of tourism can be allowed to develop.

Lesson 2: *Implementing sustainable tourism development requires measures that are both scale- and context-specific.*

Butler (Chapter 3) pointed out that sustainability (or conservation) can mean very different things in different environments and social circumstances, which makes application of the concept even more problematic. This was clearly seen in the various case studies, ranging from Place's (Chapter 9) description of the impacts of a Central American national park to Hinch's (Chapter 15) discussion of an urban cultural park, as well as the others in between. Hall (Chapter 2) and Milne (Chapter 4) both described sustainability as a contested concept that is constantly being redefined by different stakeholders. Milne further stated that sustainable tourism 'should not be viewed as an end-state, but rather as an 'ideal' towards which we can aim'. Efforts to measure or assess how far sustainable tourism efforts have gone toward that ideal (*cf.* Wight, Chapter 7), therefore, should be viewed in light of the values that the concept of 'sustainability' represents. Such values may vary, depending on the scale and situation at hand. In particular, both Milne and Butler argued for efforts to make mass tourism more sustainable, in addition to the special-interest tourism that is more closely associated with the concept. To add to this, Lew's (Chapter 8) findings indicated that trends in the ecotourism industry show an expanding mass tourism interest in remote and fragile environments, which will require stricter controls to ensure resource sustainability. Charlton (Chapter 11), however, pointed out how difficult it can be to change the travel industry and consumers to support more sustainable forms of tourism when such changes directly challenge their lifestyle – in his case car ownership and use.

Lesson 3: *Sustainable tourism issues are shaped by global economic restructuring and are fundamentally different in developing and developed economies.*

Williams and Shaw (Chapter 5), as a prelude to the industry trends noted by Lew (Chapter 8), noted that the sustainable tourism debate has closely paralleled the political economy debates on global economic restructuring. As national economies and multinational companies adapt to globalisation pressures, so too are environmental and cultural conservationists facing new challenges. Supporting Milne and Butler in their calls to make mass tourism more sustainable, Williams and Shaw felt that while sustainable development has become a global issue, much more could be done to place the sustainable tourism debate within this larger political economic context. Part of that larger context is the relationship between developed and developing economies. The historic exploitation of developing economies by developed economies continues to leave its mark on contemporary development processes. This was particularly clear in the case studies, which contrasted experiences in the USA (Chapter 10), the UK (Chapter 11), New Zealand (Chapter 14) and Canada (Chapter 15) with those in Costa Rica (Chapter 9), Nepal (Chapter 13) and among indigenous people (Chapter 6). While factors such as scale and available resources are fairly universal, the developed economies are in a better position to focus on these more traditional planning elements. Developing economies, however, have the added burden of cultural adaptation and adjustment to contend with. Lew noted that this difference can lead to misunderstandings within the travel industry, as ecotour operators from developed economies seek products for their clients that are qualitatively different from those often provided by ecotour operators in developing economies.

Lesson 4: *At the community scale, sustainable tourism requires local control of resources.*

While we may all live on one globe, most of us behave within the confines of our local communities. Zeppel (Chapter 6) and Wight (Chapter 7) both focused on the practical application of sustainable tourism at the local scale. Zeppel described tourism to indigenous communities and how these communities are attempting to preserve their identity from the direct global onslaught brought on by the processes described earlier by Williams and Shaw. She found that a sense of local control over one's resources and destiny is central to sustainable tourism development, a lesson that is appropriate to less traditional communities as well. Wight went on to provide an overview of practical assessment 'tools' that can help local communities to achieve a sense of local control. While widely used, especially for natural resource situations, she notes that none of the tools is a panacea and that they all must take into context the existing political and institutional environment. Good examples of this lesson were provided by Forbes' (Chapter 10) and Hull's (Chapter 12) descriptions of environmental tourism planning in communities that have considerable local control over their resources, and Place's (Chapter 9) example of an indigenous community that has not been able to benefit from tourism.

Lesson 5: *Sustainable tourism development requires patience, diligence, and a long-term commitment.*

Place (Chapter 9), Forbes (Chapter 10) and Charlton (Chapter 11), along with Page and Thorn (Chapter 14), described the challenges and pitfalls of developing a sustainable ecotourism industry. As can be seen in many developing areas of the world, Place showed how easily the win–win sustainable ecotourism approach can degenerate into a traditional model of exploitation of the rural poor, as short-term economic gains predominate over long-term

sustainable development. Diligence is no less of an issue in developed economies, as Forbes noted in describing the long and often very slow process of
developing an ecotourism industry in a rural American community. For a
world so accustomed to fast living and quick profits, based largely on environmentally exploitative economics, the controlled and evolutionary nature of
sustainable development can result in much scepticism and impatience. Changing these attitudes may, in the long run, prove more difficult than making
mass tourism more environmentally sensitive. Charlton, meanwhile, showed
how difficult it was to coordinate and maintain interest among numerous
stakeholders over a period of time in a venture that is only marginally successful
economically. Page and Thorn further emphasised the need for good research
and analysis to guide planning and decision making for sustainable tourism.

Geography and sustainable tourism: future directions

The contributions presented in this volume comprise a good representation of
the current state of geographic research on sustainable tourism development.
From an applied perspective, geographers are examining, and in some cases actively involved in, the sustainable development of tourism at local and regional
scales. Using the sustainable development model as a goal, they are assessing
how well communities are achieving a state of balance between resources conservation and economic exploitation. While the diversity of such experiences
is considerable, they can point the way to valuable lessons for future implementation efforts.

As is typical of academia, there is some lag between the cutting edge of
theoretical discussion and the application of theory to real-world situations. A
major evolving theoretical issue in tourism geography today appears to be that
of globalisation. This is likely to emerge more fully in future case studies as
the diverse roles of different stakeholders in a local economy are made more
explicit. Local communities are increasingly less isolated from global economic
forces, and how these global forces (e.g. multinational corporations, macro-
economic government policies and international investment markets) shape
local development is increasingly seen in newer approaches to the study of
region and place in geography.

Scale is another area in which the geography of sustainable tourism is likely
to become more explicit. As noted above, how sustainable tourism is defined
and envisioned varies by scale and circumstance. While a focus on local experiences is quite common in the geographic case study literature, more could be
done to examine sustainable tourism at the regional, national and international
scale. Different constructs of sustainable development and sustainable tourism
are likely to be necessary at these levels, offering a whole new arena for con-
testation over what these terms mean.

With regard to the distinct circumstances that each community faces,
perhaps a more appropriate model for sustainable tourism development is
one that works with a movable baseline, so that the sustainable goal can be
more clearly defined for a given community's situation. The issue of how the

baseline is identified and how the goals are defined, however, still leaves considerable opportunity for theoretical discussion and empirical research.

This book has been about sustainable tourism. We are convinced that all the contributors 'know' what sustainable tourism is within the context of their research. We are far less convinced that they could reach a consensus on what that definition is. Perhaps that is good enough. The goal of sustainable tourism development is there, just beyond our reach – and its continuing inaccessibility provides a world of opportunity for stimulating research, discussion and action.

References

Adams, R.L.A. (1966) The demand for wilderness recreation in Algonquin Provincial Park, unpublished MA thesis, Department of Geography, Clark University, Mass.

Advisory Group (1996) *1996 Performance Measures on Behalf of Fort Edmonton Park, Edmonton, Alberta,* Advisory Group, Edmonton.

Alberta Tourism Education Council (ATEC) (1993) *Freshwater Angling Guide Certification Standards,* ATEC, Edmonton.

Alderman, C. (1992) The economics and the role of privately owned lands used for nature tourism, education and conservation. Paper presented at the IVth World Parks Congress of National Parks and Protected Areas, Caracas, Venezuela.

Alger, C.F. (1988) Perceiving, analysing and coping with the local–global nexus, *International Social Science Journal* 40, 321–339.

Altman, J. (1989) Tourism dilemmas for Aboriginal Australians, *Annals of Tourism Research* 16(4), 456–476.

Altman, J. (1993) Tourism and Aboriginal communities, in B. Faulkner and M. Kennedy (eds) *Australian Tourism Outlook Forum 1992: Towards a Sustainable Tourism Future,* Bureau of Tourism Research, Canberra, 79–88.

Amin, A. (1994) *Post-Fordism: A Reader,* Blackwell, Oxford.

Ashworth, G.J. (1989) Urban tourism: an imbalance in attention, in C. Cooper (ed.) *Progress in Tourism, Recreation and Hospitality Management,* Belhaven Press, London, 33–54.

Ashworth, G.J. (1992) Is there an urban tourism? *Tourism Recreation Research* 17(2), 3–8.

Ashworth, G.J. and Tunbridge, J.E. (1990) *The Tourist-Historic City,* Belhaven Press, London.

Australia, Commonwealth of (1991) *Ecologically Sustainable Development Working Groups: Final Report – Tourism,* Australian Government Publishing Service, Canberra.

Australia, Commonwealth of (1992) *National Strategy for Ecologically Sustainable Development,* Australian Government Publishing Service, Canberra.

Bachmann, P. (1988) The Maasai – choice of East African tourists – admired and ridiculed, in P. Rossel (ed.) *Tourism: Manufacturing the Exotic,* IWGIA, Copenhagen, 47–63.

Bailey, N. (1996) Towards a research agenda for public–private partnerships in the 1990s, *Local Economy* 8(4), 292–306.

Bangs, R. (1993) Clean, green and meant to be seen: the ethos of eco-tourism, in *Proceedings of the 1993 World Congress on Adventure Travel and Eco-Tourism Manaus,* The Adventure Travel Society, 97–101.

Banister, D. (1995) Transport and the environment: a review article, *Town Planning Review* 66(4), 453–458.

Barke, M. and Newton, M. (1995) Promoting sustainable tourism in an urban context: recent developments in Malaga City, Andalusia, *Journal of Sustainable Tourism* 3(3), 115–134.

Barrett, G. (1995) Transport emissions and travel behaviour: a critical review of recent European Union and UK policy initiatives, *Transportation* 22, 295–323.

Barrow, C.J. (1995) Sustainable development: concept, value and practice, *Third World Planning Review* 17(4), 369–386.

Becker, C. (1995) Tourism and the environment, in A. Montanari and A.M. Williams (eds) *European Tourism: Regions, Spaces and Restructuring*, John Wiley, Chichester, 207–220.

Beioley, S. (1995) Green tourism – soft or sustainable? *Insights* May, B75–B89.

Belk, R.W. and Costa, J.A. (1995) International tourism: an assessment and overview, *Journal of Macromarketing* 15(2), 33–48.

Berger, D.J. (1996) The challenge of integrating Maasai tradition with tourism, in M.F. Price (ed.) *People and Tourism in Fragile Environments*, John Wiley & Sons, Chichester, 175–198.

Bernard, J. (1973) The sociology of community, in J. Reiss Jr and H.L. Wilensky (eds) *Introduction to Modern Society Series*, Scott, Foresman and Co., Glenview.

Blake, B. and Becher, A. (1997) *The New Key to Costa Rica*, Ulysses Press, Berkeley, Calif.

Blamey, R.K. and Braithwaite, V.A. (1997) A social values segmentation of the potential ecotourism market, *Journal of Sustainable Tourism* 5(1), 29–45.

Blanchard, K. (1984) Seabird harvest and the importance of education in seabird management on the North Shore of the Gulf of St Lawrence, unpublished PhD thesis, Cornell University, Ithaca, NY.

Blanchard, K. (1994) Culture and seabird conservation: the North Shore of the Gulf of St Lawrence, Canada, in D. Nettleship, J. Burger and M. Gochfeld (eds) *Seabirds on Islands: Threats, Case Studies and Action Plans*, Birdlife International Conservation, Cambridge, England, 294–310.

Blank, U. and Petkovich, M. (1987) Research on urban tourism destinations, in J.R.B. Ritchie and C.R. Goeldner (eds) *Travel, Tourism and Hospitality Research*, John Wiley & Sons, New York, 165–177.

Blunden, G., Cocklin, C. and Davis, P. (1995) Land based production in Northland, Occasional Paper 30, Department of Geography, Auckland University, Auckland.

Boo, E. (1990) *Ecotourism: The Potentials and Pitfalls*, 2 vols, World Wildlife Fund, Washington.

Boo, E. (1995) *Social Ramifications Research, Curry County Sustainable Nature-Based Tourism Project*, Curry County Commissioners, Gold Beach.

Booth, K. (1990) Recreation – a positive force for island restoration, in D.R. Towns, C.H. Daugherty and I.A.E. Atkinson (eds) *Ecological Restoration of New Zealand Islands*, Department of Conservation, Wellington, 278–283.

Bramwell, B. (1996) in W. Bramwell, I. Henry, G. Jackson, A.G. Prat, G. Richards and J. van der Straaten (eds) *Sustainable Tourism Management: Principles and Practice*, Tilburg University Press, Tilburg, 147–170.

Bramwell, B. and Lane, B. (1993) Sustainable tourism: an evolving global approach, *Journal of Sustainable Tourism* 1(1), 1–5.

Bramwell, B., Henry, I., Jackson, G., Prat, A.G., Richards, G. and van der Straaten. J. (eds) (1996) *Sustainable Tourism Management: Principles and Practice*, Tilburg University Press, Tilburg.

Brandon, K. (1996) *Ecotourism and Conservation: A Review of Key Issues*, World Bank Biodiversity Series Paper No. 33, World Bank, Washington.

Breslin, P. and Chapin, M. (1984) Conservation Kuna-style, *Grassroots Development* 8, 26–35.

Briguglio, L., Butler, R.W., Harrison, D. and Leal Filho, W. (eds) (1996) *Sustainable Tourism in Islands and Small States, Vol. II: Case Studies*, Cassell, London.

Britton, S.G. (1982) The political economy of tourism in the Third World, *Annals of Tourism Research* 9(3), 331–358.

Britton, S.G. (1991) Tourism, capital and place: towards a critical geography of tourism, *Environment and Planning D: Society and Space* 9, 451–478.

Brohman, J. (1996) New directions in tourism for Third World development, *Annals of Tourism Research* 23(1), 48–70.

Brown, L. (1981) *Building a Sustainable Society*, W.W. Norton, New York.

Bryden, J.M. (1973) *Tourism and Development: A Case Study of the Commonwealth Caribbean*, Cambridge University Press.

Buck, P.H. (1921) The evolution of the national park system of the United States, a thesis presented for the degree of Master of Arts, Ohio State University, June 1921, reprinted 1946 by the US Department of the Interior, National Park Service, United States Government Printing Office, Washington.

Buckley, R. (1994) A framework for ecotourism, *Annals of Tourism Research* 31(3), 661–665.

Burchett, C. (1992) Ecologically sustainable development and its relationship to Aboriginal tourism in the Northern Territory, in B. Weiler (ed.) *Ecotourism Incorporating the Global Classroom*, Bureau of Tourism Research, Canberra, 70–74.

Burchett, C. (1993) A profile of Aboriginal and Torres Strait Islander tourism – its history and future prospects, in *Indigenous Australians and Tourism: A Focus on Northern Australia*, AGPS, Canberra, 20–25.

Butler, R.W. (1980) The concept of a tourist area cycle of evolution: implications for management of resources, *Canadian Geographer* 24(1), 5–12.

Butler, R.W. (1989) Tourism and tourism research, in T.L. Burton and E.L. Jackson (eds) *Understanding Recreation and Leisure: Mapping the Past, Charting the Future*, Venture Publishing, State College, 567–595.

Butler, R.W. (1990) Alternative tourism: pious hope or Trojan horse?, *Journal of Travel Research* 28(3), 91–96.

Butler, R.W. (1991) Tourism, environment, and sustainable development, *Environmental Conservation* 18(3), 201–209.

Butler, R.W. (1992) Alternative tourism: the thin end of the wedge, in V. Smith and W. Eadington (eds) *Tourism Alternatives: Potentials and Problems in the Development of Tourism*, University of Pennsylvania Press, Philadelphia, 31–46.

Butler, R.W. (1993a) Pre- and post-impact assesment of tourism development, in D.G. Pearce and R.W. Butler (eds) *Tourism Research: Critiques and Challenges*, Routledge, London, 135–155.

Butler, R.W. (1993b) Tourism: an evolutionary process, in G. Nelson, R.W. Butler and G. Wall (eds) *Tourism and Sustainable Development: Monitoring, Planning, Managing*, University of Waterloo, Heritage Resources Centre, Waterloo, 27–44.

Butler, R.W., Hall, C.M. and Jenkins, J.M. (eds) (1998) *Tourism and Recreation in Rural Areas*, John Wiley, Chichester.

Butler, R.W. and Hinch, T. (eds) (1996) *Tourism and Indigenous Peoples*, International Thomson Business Press, London.

Buxton, N. (1997) The Esk Valley Line Partnership, paper to conference on 'Community–Rail Partnerships', Huddersfield, 8 April 1997.

Canada Communique (1996) Aboriginal tourism offerings partner with CTC, *Canada Communique*, November 1996.

Canadian Environmental Assessment Research Council (CEARC) (1988) *The Assessment of Cumulative Effects: A Research Prospectus*, CEARC, Ottawa.

Carr, M.H. (n.d.) Ecotourism, National Parks and Land Use Planning in Costa Rica – Llanuras of Tortuguero: A Case Study. Unpublished.

Carreras i Verdaguer, C. (1995) Mega events, local strategies and global tourist attractions, in A. Montanari and A.M. Williams (eds) *European Tourism: Regions, Spaces and Restructuring*, John Wiley, Chichester, 193–206.

Carrere, R. (1995) Loita Maasai endeavour to save sacred forest, *Shaman's Drum* 37, 17.

Carriere, J. (1991) The crisis in Costa Rica: an ecological perspective, in D. Goodman and M. Redclift (eds) *Environment and Development in Latin America*, Manchester University Press, Manchester.

Carson, R. (1962) *Silent Spring*, Fawcett, Greenwich.

Cater, E.A. (1993) Ecotourism in the third world: problems for sustainable development, *Tourism Management* 14(2), 85–90.

Cater, E.A. and Lowman, G. (eds) (1994) *Ecotourism: A Sustainable Option?* John Wiley, Chichester.

Ceballos-Lascurain, H. (1996) *Tourism, Ecotourism and Protected Areas*, IUCN, Gland.

Chang, T.C., Milne, S., Fallon, D. and Pohlmann, C. (1996) Urban heritage tourism: the global–local nexus, *Annals of Tourism Research* 23, 1–19.

Chapin, M. (1990) The silent jungle: ecotourism among the Kuna Indians of Panama, *Cultural Survival Quarterly*, 14(1), 42–45.

Charlton, C.A. (1992) *Devon and Cornwall Branch Lines Passenger Survey*, Devon and Cornwall Rail Partnership, Plymouth.

Christaller, W. (1963) Some considerations of tourism location in Europe: the peripheral regions – underdeveloped countries – recreation areas, *Papers, Regional Science Association* 12, 95–105.

Christaller, W. (1966) *Central Places in Southern Germany*, trans. C.W. Baskin, Prentice-Hall, Englewood Cliffs, NJ.

Clark, G., Darrall, J., Grove-White, R., Macnaghten, P. and Urry, J. (1994) *Leisure Landscapes – Leisure, Culture and the English Countryside: Challenges and Conflicts*, Council for the Protection of Rural England, London.

Clark, W. and Munn, R.E. (eds) (1986) *Ecologically Sustainable Development of the Biosphere*, Cambridge University Press, New York.

Clarke, J. (1997) A framework of approaches to sustainable tourism, *Journal of Sustainable Tourism* 5(3), 224–233.

Clarke, J.N. and D. McCool (1985) *Staking Out the Terrain: Power Differentials Among Natural Resource Management Agencies*, State University of New York Press, Albany.

Cloke, P., Milbourne, P. and Thomas, C. (1995) *Lifestyles in Rural England*, Rural Development Commission, London.

Coccossis, H. (1996) Tourism and sustainability: perspectives and implications, in G.K. Priestley, J.A. Edwards and H. Coccossis (eds) *Sustainable Tourism? European Experiences*, CAB International, Wallingford, 1–21.

Coccossis, H. and Nijkamp, P. (1995) *Sustainable Tourism Development*, Avebury Press, Aldershot.

Cole, D.N., Petersen, M.E. and Lucas, R.C. (1987) *Managing Wilderness Recreation Use: Common Problems and Potential Solutions*, USDA Forest Service General Technical Report INT-230, Intermountain Forest and Range Experiment Station, Utah.

Colvin, J.G. (1994) Capirona: a model of indigenous ecotourism, *Journal of Sustainable Tourism*, 2(3), 174–177.

Cook, S. (1996a) Limits of acceptable change committee meeting notes. Curry County Commissioners, Gold Beach, Oregon.

Cook, S. (1996b) Draft matrix of limits of acceptable change indicators. Curry County Commissioners, Gold Beach, Oregon.

Cornwall County Council (1996) *Transport Policies and Programmes: 1997–98 Submission*, Cornwall County Council, Truro.

Cosgrove, I. and Jackson, R. (1972) *The Geography of Recreation and Leisure*, Hutchinson, London.

Coull, C. (1996) *A Traveller's Guide to Aboriginal BC*, Beautiful British Columbia Magazine, Vancouver.

Council for the Protection of Rural England (1993) *Cornwall – Which Way Ahead? Transport Options for a Rural County*, Council for the Protection of Rural England, London.

Countryside Commission (1987) *Public Transport to the Countryside*, Countryside Commission, CCP227, Cheltenham.

Countryside Commission (1992) *Trends in Transport and the Countryside*, Countryside Commission, CCP382, Cheltenham.

Countryside Commission (1995) *Sustainable Rural Tourism: Opportunities for Local Action*, Countryside Commission, CCP483, Cheltenham.

Countryside Commission (1996a) *A Living Countryside: Our Strategy for the Next Ten Years*, Countryside Commission, CCP492, Cheltenham.

Countryside Commission (1996b) *Peace in the Parks – But too Many Cars, Survey Reveals*, Countryside Commission press release, 21 November, http://www.coi.gov.uk/depts/GCM/coi4029c.ok.

Craik, J. (1995) Are there cultural limits to tourism? *Journal of Sustainable Tourism* 3(2), 87–98.

Croall, J. (1995) *Preserve or Destroy? Tourism and the Environment*, Calouste Gulbenkian Foundation, London.

Crowe, B.L. (1969) The tragedy of the commons revisited, *Science* 166, 1103–1107.

Cullinane, S.L., Cullinane, K.P.B., Fewings, J. and Southwell, J. (1996) Rural traffic management: the Burrator Reservoir experiment, *Transport Policy* 3(4), 213–224.

Curry, N. and Pack, C. (1994) Informal county strategies for countryside recreation in England and Wales, *Journal of Environmental Management* 40, 91–101.

Daily, G. and Ehrlich, P. (1992) Population, sustainability and the earth's carrying capacity, *Bioscience* 42(10), 761–771.

Dalal-Clayton, B. (1992) Modified EIA and indicators of sustainability: first steps towards sustainability analysis, in *Industrial and Third World Environmental Assessment: The Urgent Transition to Sustainability*, the 12th Annual Meeting of the International Association for Impact Assessment, 19–22 August, World Bank, Washington, 134–146.

Daniels, A. (1995) Ahousat band's ecological walk business booms, *Vancouver Sun*, 27 January.

Dartmoor National Park Authority (1994) *Traffic Management Strategy*, Devon County Council and Dartmoor National Park Authority, Bovey Tracey.

Davies, E. (1987) Planning in the New Zealand National Parks. *New Zealand Geographer* 43(2), 73–78.

Deady, T. and Schwartzmann, M.T. (1990) Ecotourism: traveling to save the planet, *Travel Weekly* 48(84), 8–9.

Dean Runyan Associates (1994a) *Travel-Related Economic Impacts and Visitor Volume in Oregon: 1991*, Oregon Economic Development Department, Tourism Division, Salem.

Dean Runyan Associates (1994b) *Travel Industry Employment in Oregon: 1991*, Oregon Economic Development Department, Tourism Division, Salem.

Dearden, P. (1993) Cultural aspects of tourism and sustainable development: tourism and the hilltribes of Northern Thailand, in J.G. Nelson, R. Butler and G. Wall (eds) *Tourism and Sustainable Development: Monitoring, Planning and Managing*, Department of Geography, University of Waterloo, Waterloo.

Deepak, A., Regmi, R. and Rimal, N. (1994) *Nepal District Profile: A District Wide Socio-Techno-Economic Profile of Nepal*, National Research Associates, Kathmandu.

deKadt, E. (1992) Making the alternative sustainable: lessons from development for tourism, in V. Smith and W. Eadington (eds) *Tourism Alternatives: Potentials and Problems in the Development of Tourism*, Philadelphia, University of Pennsylvania Press, 47–75.

deKadt, E. (ed.) (1979) *Tourism: Passport to Development?* Oxford University Press, New York.

Department of Conservation (1994) *Visitor Strategy Discussion Document*, Department of Conservation, Wellington.

Department of the Environment (DoE) (1994) *Transport and the Environment*, Royal Commission on Environment and Pollution, HMSO, London.

Devon and Cornwall Rail Partnership (1996) *Annual Report 1995–1996*, Devon and Cornwall Rail Partnership, Plymouth.

Dex, S. (1985) *The Sexual Division of Labour*, Wheatsheaf Books, Brighton.

Dicken, P. (1994) Global–local tensions: firms and states in the global space economy, *Economic Geography* 70(2), 101–128.

Dickens, P. (1996) *Reconstructing Nature: Alienation, Emancipation and the Division of Labour*, Routledge, London.

Din, K.H. (1992) The 'involvement stage' in the evolution of a tourist destination, *Tourism and Recreation Research* 17(1), 10–20.

Ding P. and Pigram, J. (1995) Environmental audits: an emerging concept for sustainable tourism development, *Journal of Tourism Studies* 2, 2–10.

DoE (1995) *Rural England: A Nation Committed to a Living Countryside*, Department of the Environment and Ministry of Agriculture, Fisheries and Food, HMSO, London.

Dovers, S. and Handmer, J.W. (1993) Contradictions in sustainability, *Environmental Conservation* 20, 217–222.

Drake, S. (1991) Local participation in ecotourism projects, in T. Whelan (ed.) *Nature Tourism: Managing for the Environment*, Island Press, Washington, 132–163.

Dredge, D. and Moore, S. (1992) A methodology for the integration of tourism in town planning, *Journal of Tourism Studies* 3(1), 8–21.

Drummond, I. and Marsden, T.K. (1995) Regulating sustainable development, *Global Environmental Change*, 5(1), 55–63.

Duinker, P.N. (1994) Cumulative effects assessment: what's the big deal? in A.J. Kennedy (ed.) *Cumulative Effects in Canada: from Concept to Practice, Papers from the Fifteenth Symposium held by the Alberta Society of Professional Biologists*, April 13–14. Calgary, 159–178.

Dunford, M. (1990) Theories of regulation, *Society and Space* 8, 297–321.

Dyer, R. (1995) *Curry County Receives $225,000 Grant*, News Release (June), Siskiyou National Forest, Grants Pass.

Dymond, S. (1996) Application of the World Tourism Organisation's (1995) indicators of sustainable tourism within the local authority boundaries of New Zealand, unpublished Grad.Dip.Tour. dissertation, Centre for Tourism, University of Otago, Dunedin.

Eagles, P. (1992) The travel motivations of Canadian ecotourists, *Journal of Travel Research* 31(2), 3–7.

Eaton, B. and Holding, D. (1996) The evaluation of public transport alternatives to the car in British national parks, *Journal of Transport Geography* 4(1), 55–65.

Eber, S. (ed.) (1992) *Beyond the Green Horizon: Principles for Sustainable Tourism*, World Wildlife Fund, London.

Ecotourism Society (1993) *Ecotourism Guidelines for Nature Tour Operators*, ETS, North Bennington.

Editors of Rumbo (1993) Balance 'ajustado', *Rumbo* 442(25 May): 15–23.

Edmonton Journal (1991) Neighbours demand silence from fort, *Edmonton Journal*, 22 August, B2.

Eight Northern Indian Pueblos Council (1996) *8 Northern Indian Pueblos 1996 Visitors Guide*, Eight Northern Indian Pueblos Council, New Mexico.

Ekins, P. (1993a) 'Limits to growth' and 'sustainable development': grappling with ecological realities, *Ecological Economics* 8: 269–88.

Ekins, P. (1993b) Making development sustainable in W. Sachs (ed.) (1993) *Global Ecology: A New Arena of Political Conflict*, Fernwood Publications, Halifax, 91–103.

Ellis, J. (1994) Umorrduk Safaris, in *A Talent For Tourism: Stories about Indigenous People in Tourism*, Commonwealth Department of Tourism, Canberra, 31–33.

English Tourist Board, Countryside Commission and Rural Development Commission, (1991) *The Green Light: A Guide to Sustainable Tourism*, English Tourist Board, Countryside Commission and Rural Development Commission, London.

Escobar, A. (1995) *Encountering Development: The Making and Unmaking of the Third World*, Princeton University Press, Princeton, NJ.

Fairchild, H.N. (1928) *The Noble Savage: A Study in Romantic Naturalism*, Columbia University Press, New York.

FEP (1996a) *Fort Edmonton Park 1997–1999 Business Plan*, Edmonton Parks and Recreation, Edmonton.

FEP (1996b) *Criteria for Selection of a New Addition to Fort Edmonton Park* (File Copy), Edmonton Parks and Recreation, Edmonton.

FEP (1997a) *Statistical Information* (File Copy), Edmonton Parks and Recreation, Edmonton.

FEP (1997b) *Volunteer Hours Comparisons* (File Copy), Edmonton Parks and Recreation, Edmonton.

Fewings, J. (1996) A rural transport package for Dartmoor, *Town and Country Planning* June, 180–182.

Finlayson, J. (1991) *Australian Aborigines and Cultural Tourism: Case Studies of Aboriginal Involvement in the Tourist Industry*, Working Papers on Multiculturalism No. 15, The Centre for Multicultural Studies, University of Wollongong, Wollongong, Australia.

Forbes, B. (1993) *Tourism Development Plan, 1993–1998*, Brookings–Harbor Chamber of Commerce, Brookings.

Forbes, B. (1996) *Business Training Workshop Notes, July 28*, Siskiyou National Forest, Brookings, Oregon.

Foresta, R.A. (1984) *America's National Parks and Their Keepers*, Resources for the Future, Washington.

Forsberg, P., Harvey, B. and Kelsay, D. (1996) *Curry County: Public Resources From Private Initiative*, Egret Communications, Port Orford, Oregon.

Fort Edmonton Park (FEP) (1984) *Strategic Plan for Fort Edmonton Park*, Edmonton Parks and Recreation, Edmonton.

France, L. (ed.) (1997) *The Earthscan Reader in Sustainable Tourism*, Earthscan Publications, London.

Frissell, S.S. and Stankey, G.H. (1972) Wilderness environmental quality: search for social and ecological harmony. Paper presented at Annual Meeting Society of American Foresters, 4 October, Hot Springs.

Fuchs, A. and Sylvestre, J. (1995) Québec: le golfe du Saint-Laurent joue la carte du tourisme marin, *Mer et Océan* October, 48–72.

Gallie, W.B. (1955–56) Essentially contested concepts, *Proceedings of the Aristotelian Society* 56, 167–198.

Garnham, H.L. (1985) *Maintaining the Spirit of Place*, PDA Publishers, Mesa.

Gascon, C. (1994) *Discover the Lower North Shore: Guide and Directory*, Coaster's Association, St Paul's River.

Getz, D. (1986) Models in tourism planning towards integration of theory and practice, *Tourism Management* 7(1), 21–32.

Gibbs, D. (1996) Integrating sustainable development and economic restructuring: a role for regulation theory, *Geoforum* 27(1), 1–10.

Gill, R. (1994) Manyallaluk, in *A Talent For Tourism: Stories about Indigenous People in Tourism*, Commonwealth Department of Tourism, Canberra, 22–24.

Glacken, C. (1967) *Traces on the Rhodian Shore, Nature and Culture in Western Thought from Ancient Times to the End of the Eighteenth Century*, University of California Press, Berkeley.

Globe 90 (1990) *Tourism Stream: An Action Strategy for Sustainable Development*, Environment Canada, Vancouver.

Goodall, B. (1992) Environmental auditing for tourism, in C. Cooper and A.J. Lockwood (eds) *Progress in Tourism and Hospitality Management*, Wiley, Chichester, 60–74.

Goodland, R., Daly, H. and El Serafy, S. (1992) The urgent need for environmental assessment and environmental accounting for sustainability, in *Industrial and Third World Environmental Assessment: The Urgent Transition to Sustainability*, the 12th Annual Meeting of the International Association for Impact Assessment, 19–22 August, World Bank, Washington, 110–133.

Goodwin, M. and Painter, J. (1996) Local governance, the crises of Fordism and the changing geographies of regulation, *Transactions of the Institute of British Geographers* 21(4), 635–648.

Goodwin, P., Hallet, S., Kenny, F. and Stokes, G. (1991) *Transport: The New Realism*, Transport Studies Unit, University of Oxford, Oxford.

Government of Newfoundland and Labrador (1996) *Preliminary Visitor Statistics: Labrador 1995*, Government of Newfoundland and Labrador, St John's.

Graefe, A.R., Kuss, F.R. and Vaske, J.J. (1990) *Visitor Impact Management*, 2 vols., National Parks and Conservation Association, Washington.

Grano, O. (1981) External influence and internal change in the development of geography, in D.R. Stoddart (ed.) *Geography, Ideology and Social Concern*, Blackwell, Oxford, 17–36.

Gratton, C. and Straaten, J. van der (1994) The environmental impact of tourism in Europe, in C.P. Cooper and A. Lockwood (eds) *Progress in Tourism, Recreation and Hospitality Management, Vol. 5*, John Wiley, Chichester, 201–219.

Green Arrow (1997) *Green Arrow's Conservation Connection: National Eco-Agricultural Cooperative Network of Costa Rica*, http:12/10/97/www.greenarrow.com/nature/cooprena.htm.

Greenwood, D.J. (1989) Culture by the pound: an anthropological perspective on tourism as cultural commodification, in V.L. Smith (ed.) *Hosts and Guests: The Anthropology of Tourism*, University of Pennsylvania Press, Philadelphia, 171–186.

Grekin, J. and Milne, S. (1996) Toward sustainable tourism development: the case of Pond Inlet, NWT, in R.W. Butler and T. Hinch (eds) *Tourism and Native Peoples*, Routledge, London, 76–106.

Gunn, C. (1988a) *Tourism Planning*, 2nd edn, Taylor & Francis, New York.

Gunn, C. (1988b) *Vacationscape: Designing Tourist Regions*, 2nd edn, Van Nostrand Reinhold, New York.

Gunn, C.A. (1994) *Tourism Planning: Basics, Concepts, Cases*, Taylor & Francis, Washington.

Gurung, C.P. and De Coursey, M. (1994) The Annapurna Conservation Area Project: a pioneering example of sustainable tourism? in E. Cater and G. Lowman (eds) *Ecotourism: A Sustainable Option?*, John Wiley & Sons, New York, 177–194.

Hall, C.M. (1992a) *Wasteland to World Heritage: Preserving Australia's Wilderness*, Melbourne University Press, Carlton.

Hall, C.M. (1992b) *Hallmark Events: Impacts, Management and Planning*, Belhaven Press, London.

Hall, C.M. (1994a) *Tourism and Politics: Policy, Power and Place*, John Wiley, Chichester.

Hall, C.M. (1994b) Mega-events and their legacies, in P.E. Murphy (ed.) *Quality Management in Urban Tourism: Balancing Business and Environment*, University of Victoria, Victoria, 109–122.

Hall, C.M. (1995) *Introduction to Tourism in Australia: Impacts, Planning and Development*, 2nd edn, Longman Australia, South Melbourne.

Hall, C.M. (1996) Tourism and the Maori of Aotearoa, New Zealand, in R. Butler and T. Hinch (eds) *Tourism and Indigenous Peoples*, International Thomson Business Press, London, 155–175.

Hall, C.M. (1997) *Tourism in the Pacific Rim: Development, Impacts and Markets*, 2nd edn, Addison Wesley Longman, South Melbourne.

Hall, C.M. and Jenkins, J.M. (1995) *Tourism and Public Policy*, Routledge, London.

Hall, C.M., Jenkins, J.M. and Kearsley, G. (eds) (1997) *Tourism Planning and Policy in Australia and New Zealand: Cases and Issues*, Irwin Publishers, Sydney.

Hall, C.M. and Johnston, M. (eds) (1995) *Polar Tourism: Tourism in the Arctic and Antarctic Regions*, John Wiley, Chichester.

Hall, C.M., Mitchell, I. and Keelan, N. (1992) Maori cultural heritage and tourism in New Zealand, *Journal of Cultural Geography* 12(2), 115–128.

Hall, C.M. and Page, S. (eds) (1996) *Tourism in the Pacific: Cases and Issues*, International Thomson Business Press, London.

Hall, C.M. and Wouters, M.M. (1994) Managing nature tourism in the sub-Antarctic islands, *Annals of Tourism Research* 21, 355–374.

Hall, D.R. (ed.) (1991) *Tourism and Economic Development in Eastern Europe and the Soviet Union*, Belhaven Press, London.

Hall, D.R. (1995) Tourism change in Central and Eastern Europe, in A. Montanari and A.M. Williams (eds) *European Tourism: Regions, Spaces and Restructuring*, John Wiley, Chichester, 221–243.

Halliday, J. and Chehak, G. (1996) *Native Peoples of the Northwest*, Sasquatch Books, Seattle.

Hansen, N. (1992) Competition, trust and reciprocity in the development of innovative regional milieux, *Papers in Regional Science* 71(2), 95–105.

Hardin, G. (1969) The tragedy of the commons, *Science* 162, 1243–1248.

Harman, W.W. (1993) Rethinking the central institutions of modern society: society and business, *Futures* 25 (December), 1063–1071.

Harris, W.W. (1974) Three parks: an analysis of the origins and evolution of the national parks movement, unpublished MA thesis, Department of Geography, University of Canterbury, Christchurch, New Zealand.

Harrison, D. and Price, M.F. (1996) Fragile environments, fragile communities? An introduction, in M.F. Price (ed.) *People and Tourism in Fragile Environments*, John Wiley & Sons, Chichester, 1–18.

Hart, A.J. (1992) Planning for tourism in Christchurch: a comparative study, unpublished MA thesis, Department of Geography, University of Canterbury, New Zealand.

Harvey, B. and Kelsay, D. (1994–96) *Various Progress Reports – Curry County Sustainable Nature-Based Tourism Project*, Egret Communications, Port Orford, Oregon.

Harvey, B. and Kelsay, D. (1996a) *Marketing Strategy for the Curry County Sustainable Nature-Based Tourism Project*, Egret Communications, Port Orford, Oregon.

Harvey, B. and Kelsay, D. (1996b) *Target Markets for Emerging and Existing Curry County Sustainable Nature-Based Tourism Products*, Egret Communications, Port Orford, Oregon.

Hays, S.P. (1957) *The Response to Industrialism, 1885–1914*, University of Chicago Press, Chicago.

Hays, S.P. (1959) *Conservation and the Gospel of Efficiency: The Progressive Conservation Movement 1890–1920*, Harvard University Press, Cambridge, Mass.

Healy, R. (1994) 'Tourist merchandise' as a means of generating local benefits from ecotourism, *Journal of Sustainable Tourism* 2(3), 137–151.

Healy, R.G. (1994) The 'common pool' problem in tourism landscapes, *Annals of Tourism Research* 21(3), 596–611.

Healy, R.G. (1996) Tourism, Ecotourism and Sustainable Development, unpublished mimeo, Duke University, Raleigh, NC.

Hecht, S., Anderson, A. and May, P. (1988) The subsidy from nature: shifting cultivation, successional palm forests and rural revelopment, *Human Organization* 47(1), 25–35.

Heeley, J. (1981) Planning for tourism in Britain, *Town Planning Review* 52, 61–79.

Hendee, J.C., Stankey, G.H. and Lucas, R.C. (1978) *Wilderness Management*, Miscellaneous Publications No. 1365 USDA Forest Service, Washington.

Hill, C. (1990) The paradox of tourism in Costa Rica, *Cultural Survival Quarterly* 14(1), 14–19.

Hinch, T.D. (1996) Urban tourism: perspectives on sustainability, *Journal of Sustainable Tourism* 4(2), 95–110.

Hinch, T.D. and Butler, R.W. (1996) Indigenous tourism: a common ground for discussion, in R. Butler and T. Hinch (eds) *Tourism and Indigenous Peoples*, International Thomson Business Press, London, 3–19.

HMGN: Ministry of Tourism and Civil Aviation Department of Tourism (1994) *Nepal Tourism Statistics 1993*, Rising Sun Printers, Kathmandu.

Hof, M., Hammett, J., Rees, M., Beinap, J., Poe, N., Lime, D. and Manning, B. (1994) Getting a handle on visitor carrying capacity – a pilot project at Arches National Park, *Park Science* Winter, 11–13.

Honey, M. (1994) Paying the price of ecotourism, *Americas* 46(6), 40–47.

Honour, H. (1975) *The New Golden Land: European Images of America from the Discoverers to the Present Times*, Pantheon Books, New York.

Honour, H. (1981) *Romanticism*, Penguin Press, Harmondsworth.

House of Commons Environment Committee (1995) *The Environmental Impact of Leisure Activities*, HMSO, London.

Howard, C. (1993) Costa Rica 'selling out' to foreigners, *San Francisco Chronicle* 9 August, A10.

Howe, J. (1982) Kindling self-determination among the Kuna, *Cultural Survival Quarterly* 6(3), 15–17.

Hudson, R. (1995) Towards sustainable industrial production: but in what sense sustainable? in M. Taylor (ed.) *Environmental Change: Industry, Power and Policy,* Avebury, Aldershot, 37–56.

Hudson, R. and Williams, A.M. (1994) *Divided Britain,* Wiley, Chichester.

Hull, J.S. (1996) Using the Internet to promote a sustainable tourism industry: a case study of the Lower North Shore of Quebec. Paper presented at the International Geographical Union. The Hague, Netherlands. August 1996.

Hutchinson, J. (1994) The practice of partnership in local economic development, *Local Government Studies* 20(3), 335–344.

Iaonnides, D. (1995) Strengthening the ties between tourism and economic geography: a theoretical agenda, *Professional Geographer* 47(1), 49–60.

Inskeep, E. (1991) *Tourism Planning: An Integrated and Sustainable Development Approach,* Van Nostrand Reinhold, New York.

International Institute for Sustainable Development, Deloitte and Touche and The Business Council for Sustainable Development (1992) *Business Strategy for Sustainable Development, Leadership and Accountability in the '90s,* IISD, Winnipeg.

International Union for the Conservation of Nature and Natural Resources (IUCN) (1980) *World Conservation Strategy,* The IUCN with the advice, cooperation and financial assistance of the United Nations Environment Education Programme and the World Wildlife Fund and in collaboration with the Food and Agriculture Organisation of the United Nations and the United Nations Educational, Scientific and Cultural Organisation, IUCN, Morges.

Ise, J. (1961) *Our National Park Policy A Critical History,* published for Resources for the Future by The Johns Hopkins Press, Baltimore, Md.

Jackson, C. (1994) Gender analysis and environmentalism, in M. Redclift and T. Benton (eds) *Social Theory and the Global Environment,* Routledge, London, 113–149.

Jackson, P. (1995) Changing geographies of consumption, *Environment and Planning A* 27, 1875–1876.

Jackson, P. and Thrift, N. (1995) Geographies of consumption, in D. Miller (ed.) *Acknowledging Consumption,* Routledge, London, 204–237.

Jacob. M. (1994) Toward a methodological critique of sustainable development, *Journal of Developing Areas* 28, 237–252.

Jamal, T.B. and Getz, D. (1995) Collaboration theory and community tourism planning, *Annals of Tourism Research* 22(1), 186–204.

Jamieson, W. (ed.) (1990) *Maintaining and Enhancing the Sense of Place for Small Communities and Regions,* University of Calgary, Calgary.

Jansen-Verbeke, M. (1986) Inner city tourism: resources, tourists, and promoters, *Annals of Tourism Research* 13(1), 79–100.

Jefferson, T. (1861) *Notes on the State of Virginia,* Harper & Row, New York.

Jessop, B., Peck, J. and Tickell, A. (1996) Retooling the machine: economic crisis, state restructuring, and urban politics. Paper presented to the Annual Meeting of the Association of American Geographers, 9–13 April, Charlotte.

Johnston, C.R. (ed.) (1990) Breaking out of the tourist trap, *Cultural Survival Quarterly* 14(1 & 2).

Johnston, R.J. (1991) *Geography and Geographers: Anglo-American Human Geography Since 1945,* 4th edn, Edward Arnold, London.

Jones, G. (1991) Le développement touristique de la zone Vieux-fort/Blanc-Sablon en Basse-Côte-Nord du Golfe St Laurent, unpublished Master's thesis, Université du Québec à Rimouski, Rimouski, Québec.

Jones, P. (1993) Wildlife viewing in Atlantic Canada: A case study of constraints and opportunities, unpublished Master's project, Simon Fraser University, Burnaby.

Jones, S. (1995) *Walk the Wild Side*, facsimile to Canadian National Aboriginal Tourism Association British Columbia Region, 15 March 1995.

Jones, S.B. (1933) Mining tourist towns in the Canadian Rockies, *Economic Geography* 9, 368–378.

Kay, J. and Schneider, E. (1994) Embracing complexity: the challenge of the ecosystem approach, *Alternatives* 20(3): 32–39.

Keelan, N. (1996) Maori heritage: visitor management and interpretation, in C.M. Hall and S. McArthur (eds) *Heritage Management in Australia and New Zealand: The Human Dimension*, Oxford University Press, Melbourne, 195–201.

King Mahendra Trust for Nature Conservation (KMTNC) (1994a) *Annapurna Conservation Area Project: Conservation for Development*, KMTNC, Kathmandu.

KMTNC (1994b) *King Mahendra Trust For Nature Conservation: A Decade of Conservation and Development (1984–1994)*, KMTNC, Kathmandu.

Kinnaird, V. and Hall, D. (eds) (1994) *Tourism: A Gender Analysis*, John Wiley, Chichester.

Kirkby, S.J. (1995) Introduction, in J. Kirkby, P. O'Keefe, and L. Timberlake (eds) *The Earthscan Reader in Sustainable Development*, Earthscan Publications, London.

KPMG Consulting (1995) *Technology and Tourism Marketing in Canada*, Canadian Tourism Commission, Ottawa.

Kramer, P. (1994) *Native Sites in Western Canada*, Altitude Publishing, Canmore.

Krawetz, N.M., MacDonald, W.R. and Nichols, P. (1987) *A Framework for Effective Monitoring*, Canadian Environmental Assessment Research Council, Ottawa.

Kretchmann, J. and Eagles, P. (1990) An analysis of the motives of ecotourists in comparison to the general Canadian population, *Society and Leisure* 13(2): 499–507.

Laarman, J.G. and Durst, P.B. (1987) *Nature Travel in the Tropics*, FPEI Working Paper No. 23. Southeastern Center for Forest Economics Research, Research Training Park, North Carolina.

Laarman, J.G. and Perdue, R. (1989) Science tourism in Costa Rica, *Annals of Tourism Research* 16, 205–215.

Lang, R. (1988) Planning for integrated development, in F.W. Dykeman (ed.) *Integrated Rural Planning and Development*, Mount Allison University, Sackville, 81–104.

Larsen, K (1993) Around Manaslu, *Buzzworm: The Environmental Journal* 5, 60–64.

Lash, S. and Urry, J. (1994) *Economies of Signs and Spaces*, Sage, London.

Lavery, P. (ed.) (1971) *Recreational Geography*, David and Charles, London.

Law, C.M. (1993) *Urban Tourism: Attracting Visitors to Large Cities*, Mansell, London.

Leininger, A. (1993) *Algunos aspectos de la geografia turistica y su aplicacion en Costa Rica*, Escuela de Ciencias Geograficas, Universidad Nacional Costa Rica, Heredia, unpublished.

Leiper, N. (1990) Tourist attraction systems, *Annals of Tourism Research* 17(3), 367–384.

Lemons, J. and Brown, D.A. (1995) *Sustainable Development: Science, Ethics, and Public Policy*, Kluwer Academic Publishers, London.

Lew, A.A. (1987) A framework of tourist attraction research, *Annals of Tourism Research* 14(3), 553–575.

Lew, A.A. and van Otten, G.A. (eds) (1998) *Tourism and Gaming on American Indian Lands*, Cognizant Communications Corporation, New York.

Lew, A.A. and Yu, L. (eds) (1995) *Tourism in China: Geographic, Political and Economic Perspectives*, Westview Press, Boulder, Colorado.

Lime, D.W. and Stankey, G.H. (1971) Carrying capacity: maintaining outdoor recreation quality, in *Forest Recreation Symposium Proceedings*, Northeast Forest Experimental Station, Upper Darby, 174–184.

Lindberg, K. (1991) *Policies for Maximizing Nature Tourism's Ecological and Environmental Benefits*, World Resources Institute, Washington.

Lipietz, A. (1992) *Towards a New Economic Order: Postfordism, Ecology and Democracy*, Polity Press, Cambridge.

Long, V.H. (1990) Tourism development, conservation and anthropology, *Practising Anthropology* 14(2), 14–17.

Lowenthal, D. (1958) *George Perkins Marsh: Versatile Vermonter*, Columbia University Press, New York.

Lowndes, V., Nanton, P., McCabe, A. and Skelcher, C. (1997) Networks, partnerships and urban regeneration, *Local Economy* 11(4), 333–342.

Lowyck, E., Van Langenhove, L. and Bollaert, L. (1992) Typologies of tourist roles, in P. Johnson and T. Barry (eds), *Choice and Demand in Tourism*, Mansell, London, 13–32.

Lujan, C.C. (1993) A sociological view of tourism in an American Indian community: maintaining cultural integrity at Taos Pueblo, *American Indian Culture and Research Journal* 17(3), 101–120.

Lumsden, L. (1992) The Peak District National Park: The Upper Derwent, *Marketing for Tourism: Case Study Assignments*, Macmillan, London.

MacCannell, D. (1976) *The Tourist: A New Theory of the Leisure Class*, Schoken Books, New York.

MAFF (1994) *Success with Sporting Enterprises on Farms: A Guide to Farmers and Landowners*, Ministry of Agriculture, Fisheries and Food, London.

Maiden, A.N. (1995) Exploring the Maori experience, *The Independent Monthly*, July, 74–81.

Mak, A. (1981) The Montagnais, In *The Lower North Shore*. Ministère des Affaires Culturelles du Québec, Québec, 86–99.

Mallari, A.A. and Enote, J.A. (1996) Maintaining control: culture and tourism in the Pueblo of Zuni, New Mexico, in M.F. Price (ed.) *People and Tourism in Fragile Environments*, John Wiley & Sons, Chichester, 19–31.

Manidis Roberts Consultants (1995) *Determining an Environmental and Social Carrying Capacity for Jenolan Caves Reserve*, Manidis Roberts, Sydney.

Marsh, G.P. (1965) *Man and Nature or, Physical Geography as Modified by Human Action*, orig. 1864, ed. D. Lowenthal, The Belknap Press of Harvard University Press, Cambridge, Mass.

Marsh, G.P. (1968) Man's responsibility for the land, in R. Nash (ed.) *The American Environment, Readings in the History of Conservation*, Addison-Wesley Publishing, Reading, Mass., 13–18.

Mathieson, A. and Wall, G. (1982) *Tourism: Economic, Physical and Social Impacts*, Longman, Harlow.

Matley, I.M. (1976) *The Geography of International Tourism*, Resource Paper 76–1, Association of American Geographers, Washington.

Matthews, W.H. (1975) Objective and subjective judgement in environmental impact analysis, *Environment Conservation* 2, 121–131.

Mattson, V.E. (1985) *Frederick Jackson Turner: A Reference Guide*, G.K. Hall, Boston.

McCool, S.F. (1991) Limits of acceptable change: a strategy for managing the effects of nature-dependent tourism development, Paper presented at Tourism and the Land: Building a Common Future, 1–3 December 1990, Whistler.

McCool, S.F. (1994) Planning for sustainable nature-dependent tourism development: the limits of acceptable change system, *Tourism Recreation Research* 19(2), 51–55.

McCool, S.F. and Stankey, G.H. (1992) Managing for the sustainable use of protected wildlands: the limits of acceptable change framework. Paper presented at IVth World Congress on National Parks and Protected Areas, Caracas, Venezuela, February 10–21.

McHugh, E. (1996) Time travelling, in *The Australian Magazine, Territory Tapestry: Special Promotion for the Australian Magazine*, 23–24 November, 35–36.

McIntyre, G. (1993) *Sustainable Tourism Development: Guide for Local Planners*, World Tourism Organisation, Madrid.

McKercher, B. (1993a) Some fundamental truths about tourism: understanding tourism's social and environmental impacts, *Journal of Sustainable Tourism* 1(1), 6–16.

McKercher, B. (1993b) The unrecognized threat to tourism: can tourism survive sustainability? *Tourism Management* 14(2), 131–136.

McMurray, K.C. (1930) The use of land for recreation, *Annals of the Association of American Geographers* 20, 7–20.

McPhedran, K. (1995) Long Beach: the wetter the better, *Traveller* 4(3), 6–13.

Meadows, D.H., Meadows, D.L., Randers, J. and Behrens, W.W. (1972) *Limits to Growth*, Universal Books, New York.

Mercer, D.C. (1970) The geography of leisure: a contemporary growth point, *Geography* 55(3), 261–273.

Mercer, D.C. (1994) Native peoples and tourism: conflict and compromise, in W.F. Theobald (ed.) *Global Tourism: The Next Decade*, Butterworth-Heinemann, Boston, 124–145.

Metz, J.J. (1991) A reassessment of the causes and severity of Nepal's environmental crisis, *World Development* 19, 805–820.

Mikesell, R.F. (1992) Environmental assessment and sustainability at the project and program level, in *Industrial and Third World Environmental Assessment: The Urgent Transition to Sustainability*, the 12th Annual Meeting of the International Association for Impact Assessment, 19–22 August, World Bank, Washington, 86–92.

Milne, S. (1997) Tourism, dependency and South Pacific microstates: beyond the vicious cycle? in D.G. Lockhart and D. Drakakis-Smith (eds) *Island Tourism: Trends and Prospects*, Pinter, London, 281–301.

Milne, S. and Gill, K. (forthcoming) Distribution technologies and destination development: myths and realities, in K. Debbage and D. Iaonnides (eds) *The Economic Geography of Tourism*, Routledge, London.

Milne, S. and Pohlmann, C. (forthcoming) The evolving hotel industry: evidence from Montreal, in K. Debbage and D. Iaonnides (eds) *The Economic Geography of Tourism*, Routledge, London.

Milne, S. and Wenzel, G. (1991) Tourism and economic development in the Baffin Region of Canada's Northwest Territories, unpublished paper, McGill University, Montreal.

Milne, S., Tarbotton, R., Woodley, S. and Wenzel, G. (1997) *Tourists to the Baffin Region: 1992 and 1993 Profiles*, McGill Tourism Research Group Industry Report No. 11, McGill University, Montreal.

Ministère des Transports du Québec (1988) *Statistiques de transports maritimes*, Gouvernement du Québec, Québec.

Ministry of Tourism (1992) *Tourism Sustainability*, Discussion Paper, Ministry of Tourism, Wellington.

Ministry of Tourism (1993) *The Resource Management Act – A Guide for the Tourism Industry*, Ministry of Tourism, Wellington.

Mitchell, B. (1989) *Geography and Resource Analysis*, 2nd edn, Longman, Harlow.

Mitchell, L. (1969) Recreational geography: evolution and research needs, *Professional Geographer* 21(2), 117–119.

Mitchell, L. and Murphy, P. (1991) Geography and tourism, *Annals of Tourism Research* 18, 57–70.

Montanari, A. and Muscara, C. (1995) Evaluating tourist flows in historic cities: the case of historic cities, *Tijdschrift voor Economische en Sociale Geografie* 86, 80–88.

Montanari, A. and Williams, A.M. (1995) *European Tourism: Regions, Spaces and Restructuring*, Wiley, Chichester.

Mora Castellano, E. (1996) *El secuestro de una nacion, Ambien-Tico* (March 1996), Revista mensual del proyecto 'Actualidad Ambiental en Costa Rica', Escuela de Ciencias Ambientales, Universidad Nacional de Costa Rica, No. 38, 9–11.

Mowforth, M. (1996) *Rural Rail Branch Lines as Axes of Economic Regeneration*, Devon and Cornwall Rail Partnership, Plymouth.

Mowforth, M. and Charlton, C.A. (1996) *Valuing Rural Rail Branch Lines: A Methodology for Investigation and Guide for Potential Researchers*, Devon and Cornwall Rail Partnership, Plymouth.

Mowforth, M. and Munt, I. (1997) *Tourism and Sustainability*, Routledge, London.

Mulchand, S. (1993) Northern exposure, *Asia Travel Trade*, September, 26–27.

Muller, H. (1997) The thorny path to sustainable tourism, in L. France (ed.) *The Earthscan Reader in Sustainable Tourism*, Earthscan Publications, London, 29–35.

Mullings, B. (1996) Globalization, economic restructuring and the development process in Jamaica: the case of telecommunications, unpublished PhD thesis, Department of Geography, McGill University, Montreal.

Munro, D.A. (1986) Environmental impact assessment as an element of environmental management, in the Canadian Environmental Assessment Research Council and the US National Research Council Board on Basic Biology (eds) *Proceedings of the Workshop on Cumulative Environmental Effects: A Binational Perspective*, Federal Environmental Assessment Review Office, Hull, 25–30.

Muqbil, I. (1996) Growth highlights tourism quandary, *Travel News Asia*, 27 October, http://web3.asia1.com.sg/timesnet/data/tna/docs/tna3537.html.

Murphy, P.E. (1985) *Tourism: A Community Approach*, Methuen, New York.

Murphy, P.E. (1994) Tourism and sustainable development, in W.F. Theobald (ed.) *Global Tourism: The Next Decade*, Butterworth-Heinemann, Oxford, 274–290.

Murphy, R.E. (1963) Geography and outdoor recreation: an opportunity and an obligation, *Professional Geographer* 15(5), 33–34.

Myers, N. and Gaia Ltd staff (1984) *Gaia: An Atlas of Planet Management*, Anchor/Doubleday, New York.

Nash, R. (1963) The American wilderness in historical perspective, *Journal of Forest History* 6(4), 2–13.

Nash, R. (1967) *Wilderness and the American Mind*, Yale University Press, New Haven, Conn., and London.

Nash, R. (ed.) (1968) *The American Environment: Readings in the History of Conservation*, Addison-Wesley Publishing, Reading, Mass.

Nath, B., Hens, L. and Devuyst, D. (eds) (1996) *Sustainable Development*, VUB University Press, Brussels.

Nathan, H. (1991) Heritage centre owners to offer creditors buyout, *Times Colomnist*, 6 April.

Nelson, J.G. (1973) Canada's national parks – past, present and future, *Canadian Geographical Journal* 86(3), 68–89.

Nelson, J.G. (1993) An introduction to tourism and sustainable development with special reference to monitoring, in J.G. Nelson, R. Butler and G. Wall (eds) *Tourism and Sustainable Development: Monitoring, Planning, Managing*, Department of Geography, University of Waterloo, Waterloo, Ontario.

Nelson, J.G. and Butler, R.W. (1975) Recreation and the environment, in I.R. Manners and M.W. Mikesell (eds) *Perspectives on Environment*, Association of American Geographers, Washington, 290–310.

Nelson, J.G., Butler, R. and Wall, G. (eds) (1993) *Tourism and Sustainable Development: Monitoring, Planning, Managing*, Department of Geography Publication Series No. 37, Heritage Resource Centre, University of Waterloo, Waterloo, Ontario.

New Zealand (1887) *Parliamentary Debates* 57, 399.

New Zealand (1894) *Parliamentary Debates* 86, 579.

New Zealand Tourism and Publicity Department (NZTPD) (1988) *The Implications of Tourism Growth in New Zealand*, NZTPD, Wellington.

New Zealand Tourism Board (NZTB) (1991) *Tourism in the 1990s*, NZTB, Wellington.

Nga Korero (1995a) Tourism, *Nga Korero*, February.

Nga Korero (1995b) Tourism, *Nga Korero*, May.

Nga Korero (1996a) Resources, *Nga Korero*, November.

Nga Korero (1996b) Tourism, *Nga Korero*, January.

Nicholson, M.H. (1962) *Mountain Gloom and Mountain Glory*, Norton, New York.

Nicholson-Lord, D. (1994) Hitting – and obliterating – the trail, *World Press Review* January, 48.

Nicoara, A. (1992) Ecotourism: Trojan horse or savior of the last, best place? *Travel Matters (Moon Publications Alternative Travel Newsletter)* 4.

Northern Territory Tourist Commission (NTTC) (1994) *Aboriginal Tourism in the Northern Territory: A Discussion Paper*, NTTC, Darwin.

Nove, J. (1995) *Quebec Lower North Shore Visitor's Guide*, QLF/Atlantic Centre for the Environment, Ipswich, Mass.

NTTC (1996) *Aboriginal Tourism Strategy*, NTTC, Darwin.

NZTB (1993) *New Zealand International Visitor Survey 1992/93*, NZTB, Wellington.

NZTB (1994) *Tourism Investment and the Resource Management Act 1991 – Discussion Document*, NZTB, Wellington.

O'Neil, J. (1996) *Ladicte coste du nord*, Libre Expression, Montreal.

O'Reilly (1986) Tourism carrying capacity: concepts and issues, *Tourism Management* 7(4), 254–258.

Office of Passenger Rail Franchising (Opraf) (1996) New benefits for passengers in South Wales and West and Cardiff Railway franchise awards, *Opraf News Release*, 17 September, 1–10.

Olmsted, F.L. and Wharton, W.P. (1932) The Florida Everglades: where the mangrove forests meet the storm waves of a thousand miles of water, *American Forests* 38 (March), 142–147, 192.

Olwig, K. and Olwig, K. (1979) Underdevelopment and the development of 'natural' parks ideology, *Antipode* 11(2), 16–25.

Page, S.J. (1995) *Urban Tourism*, Routledge, London.

Page, S.J. and Hall, C.M. (1998) *The Geography of Tourism and Recreation*, Routledge, London.

Page, S.J. and Piotrowski, S. (1990) A critical evaluation of international tourism in New Zealand, *British Review of New Zealand Studies* 3, 87–108.

Pannell Kerr Forster Management Consultants (1990) *Fort Edmonton Park: Opportunities for a Major Tourist Attraction*, Pannell Kerr Forster Management Consultants, Edmonton.

Parks Canada (1994) *Parks Canada – Guiding Principles and Operating Policies*. Ministry of Supply and Services, Ottawa.

Pearce, D. (1988) Economics, equity and sustainable development, *Futures* 20, 598–605.

Pearce, D., Barbier, E. and Markandy, A. (1987) *Sustainable Development and Cost–Benefit Analysis*, London Environmental Economics Centre, Paper 88-01.

Pearce, D.G. (1979) Towards a geography of tourism, *Annals of Tourism Research* 6(3), 245–272.

Pearce, D.G. (1985) Tourism and planning in the Southern Alps of New Zealand, in T. Singh and J. Kaur (eds) *Integrated Mountain Development*, Himalayan, New Delhi.

Pearce, D.G. (1989) *Tourist Development*, 2nd edn, Longman, Harlow.

Pearce, D.G. (1990) Tourism, the regions and restructuring in New Zealand, *Journal of Tourism Studies* 1(2), 33–42.

Pearce, D.G. (1991) *Tourism Today: A Geographical Analysis*, Longman, Harlow.

Pearce, D.G. (1992) *Tourist Organisations*, Longman, Harlow.

Pearce, D.G. (1995) Planning for tourism in the 1990s: An integrated, dynamic multiscale approach, in R. Butler and D.G. Pearce (eds) *Change in Tourism: People, Places and Processes*, Routledge, London, 229–244.

Pearce, P.L. (1991) Analyzing tourist attractions, *Journal of Tourism Studies* 2(1), 46–55.

Pearce, P.L. (1993) Fundamentals of tourist motivation, in D.G. Pearce and R.W. Butler (eds) *Tourism Research: Critiques and Challenges*, Routledge, London, 113–134.

Pearce, P.L. (1995) From cultural shock and culture arrogance to culture exchange: ideas towards sustainable socio-cultural tourism, *Journal of Sustainable Tourism* 3(3), 143–154.

Peck, J. and Tickell, A. (1992) Local modes of social regulation?: regulation theory, Thatcherism and uneven development, *Geoforum* 23(3), 347–363.

Pepper, D. (1984) *The Roots of Modern Environmentalism*, Croom Helm, London.

Perkins, H., Devlin, P., Simmons, D. and Batty, R. (1993) Recreation and tourism, in A. Memon and H. Perkins (eds) *Environmental Planning in New Zealand*, Dunmore Press, Palmerston North, 169–192.

Perrottet, T. (1996) Grace & fervour, *The Weekend Australian, The Weekend Review: Travel*, 14–15 December, 16.

Pigram, J. (1983) *Outdoor Recreation Resource Management*. Croom Helm, Kent.

Pigram, J.J. (1990) Sustainable tourism: policy considerations, *Journal of Tourism Studies*, 1(2), 2–9.

Pinchot, G. (1968) Ends and means, in R. Nash (ed.) *The American Environment: Readings in the History of Conservation*, Addison-Wesley, Reading, Mass., 59–64.

Pizam, A. and Knowles, T. (1994) The European hotel industry, in C.P. Cooper and A. Lockwood (eds) *Progress in Tourism, Recreation and Hospitality Management, Vol. 5*, John Wiley, Chichester, 135–152.

Place, S.E. (1988) The impact of national park development on Tortuguero, Costa Rica, *Journal of Cultural Geography* 9(1), 37–52.

Place, S.E. (1991) Nature tourism and rural development in Tortuguero, *Annals of Tourism Research* 18, 186–201.

Place, S.E. (1995) Ecotourism for sustainable development: oxymoron or plausible strategy? *GeoJournal* 35(2), 161–174.

Poon, A. (1990) Flexible specialization and small size: the case of Caribbean tourism, *World Development* 18(1), 109–123.

Poon, A. (1993) *Tourism, Technology and Competitive Strategies*, CAB International, Wallingford.

Poon, A. (1994) The 'new tourism' revolution, *Tourism Management* 15(2), 91–92.

Potvin, J.R. (1991) *Indicators of Ecologically Sustainable Development: Synthesized Workshop Proceedings*, Canadian Environmental Advisory Council, Ottawa.

Powell, J.M. (1976) *Conservation and Resource Management in Australia 1788–1914, Guardians, Improvers and Profit: An Introductory Survey*, Oxford University Press, Melbourne.

Price, M.F. (ed.) (1996) *People and Tourism in Fragile Environments*, John Wiley & Sons, Chichester.

Pridham, G. (1996) *Tourism Policy in Mediterranean Europe: Towards Sustainable Development?* Occasional Paper 15, University of Bristol, Centre for Mediterranean Studies, Bristol.

Priestley, G.K., Edwards, J.A. and Coccossis, H. (1996) *Sustainable Tourism? European Experiences*, CAB International, Wallingford.

Prior, D. (1996) 'Working the network': local authority strategies in the reticulated local state, *Local Government Studies* 22(2), 92–104.

RDC (1997) *The Economic Impact of Recreation and Tourism in the English Countryside*, Rural Development Commission and Countryside Commission, London.

Redclift, M. (1987) *Sustainable Development: Exploring the Contradictions*, Methuen, London.

Redclift, M. (1988) Sustainable development and the market: a framework for analysis, *Futures* 20(6), 638.

Redclift, M. (1995a) Sustainable development and popular participation: a framework for analysis, in D. Ghai and J. Vivian (eds) *Grassroots Environmental Action: People's Participation in Sustainable Development*, Routledge, London, 23–49.

Redclift, M. (1995b) Sustainability and theory: some points of departure. Paper presented at the Workshop on the Political Economy of the Agro Food System, Institute of International Studies, Berkeley, Calif., September 28–30.

Redclift, M. and Benton, T. (1994) *Social Theory and the Global Environment*, Routledge, London.

Rees, W.E. (1990) Economics, ecology, and the role of environmental assessment in achieving sustainable development, in P. Jacobs and B. Sadler (eds) *Sustainable Development and Environmental Assessment: Perspectives on Planning for a Common Future*, Canadian Environmental Research Council, Hull.

Rees, W.E. (1994) Personal communication.

Rees, W.E. and Wackernagel, M. (1994) Ecological footprints and appropriated carrying capacity: measuring the natural capital requirements of the human economy, in A.-M. Jansson, M. Hammer, C. Folke and R. Costanza (eds) *Investing in Natural Capital*, Island Press, Washington.

Relph, E. (1976) *Place and Placelessness*, Pion, London.

Resource Assessment Commission (1993) *The Carrying Capacity Concept and its Application to the Management of Coastal Zone Resources*, Information Paper No. 8. AGPS, Canberra.

Resource Management Act (1991) *Resource Management Act 1991*, Government Print, Wellington.

Rhodes, C. (1986) Management perspective: commentary I, in *Proceedings of the Workshop On Cumulative Environmental Effects: A Binational Perspective*, the Canadian Environmental Assessment Research Council and the US National Research Council Board on Basic Biology, Ottawa, 31–33.

Richardson, E.R. (1962) *The Politics of Conservation: Crusades and Controversies 1897– 1913*, University of California Press, Berkeley and Los Angeles.

Roberts, J. (1996) Green consumers in the 1990s: profile and implications for advertising, *Journal of Business Research*, 36, 217–231.

Robinson, G. (1993) Tourism and tourism policy in the European Community: an overview, *International Journal of Hospitality Management* 12, 7–20.

Robinson, H. (1976) *A Geography of Tourism*, MacDonald and Evans, London.

Robinson, J.B., Francis, G., Legge, R. and Lerner, S. (1990) Defining a sustainable society: values, principles and definitions, *Alternatives: Perspectives on Society, Technology and Environment* 17(2), 36–46.

Rooke, B. (1993) Finance – the Umorrduk experience, in *Indigenous Australians and Tourism: A Focus on Northern Australia*, AGPS, Canberra, 38–39.

Root, A., Boardman, B. and Fielding, W.J. (1996) *The Costs of Rural Travel*, Environmental Change Unit, University of Oxford, Oxford.

Ross, W.A. (1994) Assessing cumulative environmental effects: both impossible and essential, in A.J. Kennedy (ed.) *Cumulative Effects in Canada: From Concept to Practice*, Papers from the Fifteenth Symposium held by the Alberta Society of Professional Biologists, April 13, 14. Calgary, 1–9.

Rudkin, B. and Hall, C.M. (1996) Unable to see the forest for the trees: ecotourism development in Solomon Islands, in R. Butler and T. Hinch (eds) *Tourism and Indigenous Peoples*, 2nd edn, International Thomson Business Press, London, 203–226.

Ruitenbeek, H.J. (1991) *Indicators of Ecologically Sustainable Development: Towards New Fundamentals*, Canadian Environmental Advisory Council, Ottawa.

Runte, A. (1974) Yosemite Valley Railroad highway of history, *National Parks and Conservation Magazine: The Environmental Journal*, 48 (December), 4–9.

Runte, A. (1979) *National Parks The American Experience*, University of Nebraska Press, Lincoln.

Rural Development Commission (RDC) (1992) *Harvesting the Benefits from Visitors to the Countryside*, Rural Development Commission, London.

Russsell, B. (1946) *A History of Western Philosophy*, Counterpoint Unwin Paperbacks, London.

Ryel, R. and Grasse, T. (1991) Marketing ecotourism: attracting the elusive ecotourist, in T. Whelan (ed.) *Nature Tourism*, Island Press, Washington, 164–186.

Sachs, W. (ed.) (1993) *Global Ecology: A New Arena of Political Conflict*, Fernwood Publications, Halifax, NovaScotia.

Sadler, B. (1986). Commentary II, in *Proceedings of the Workshop On Cumulative Environmental Effects: A Binational Perspective*, the Canadian Environmental Assessment Research Council and the US National Research Council Board on Basic Biology, Ottawa, 71–75.

Sax, J.L. (1976) America's national parks: their principles, purposes and prospects, *Natural History* 85(8), 57–88.

Sayer, A. and Walker, R. (1992) *The New Social Economy*, Blackwell, Oxford.

Schaller, D.T. (1996) *Indigenous Ecotourism and Sustainable Development: The Case of Rio Blanco, Ecuador*, http:// . . . ww.geog.umn.edu/~schaller/RioBlancoSummary. html.

Schneider, D.M., Godschalk, D.R. and Axler, N. (1978) *The Carrying Capacity Concept as a Planning Tool*, Planning Advisory Service Report No. 338, American Planning Association, Chicago.

Schwepker, C. and Cornwell, T. (1991) An examination of ecologically concerned consumers and their intention to purchase ecologically packaged products, *Journal of Public Policy and Marketing* 10(2), 77–101.

Scottish Tourist Board (1994) *Tourism and the Environment*, Scottish Tourist Board, Edinburgh.

Shackley, M. (1994) The land of Lo, Nepal/Tibet: the first eight months of tourism, *Tourism Management* 15, 17–26.

Shankland, R. (1970) *Steve Mather of the National Parks*, 3rd edn, Alfred A. Knopf, New York.

Shaw, G. and Williams, A.M. (1994) *Critical Issues in Tourism: A Geographical Perspective*, Blackwell, Oxford.

Shelby, B. and Heberlein, T.A. (1984) A conceptual framework for carrying capacity determination, *Leisure Sciences* 6(4), 433–451.

Smith, S.L. (1989) *Tourism Analysis*, Longman, Harlow.

Smith, V. (ed.) (1977) *Hosts and Guests: The Anthropology of Tourism*, University of Pennsylvania Press, Philadelphia.

Smith, V. (1989) Eskimo tourism: micro-models and marginal men, in V.L. Smith (ed.) *Hosts and Guests: The Anthropology of Tourism*, 2nd edn, University of Pennsylvania Press, Philadelphia, 55–82.

Smith, V. (1996) The Inuit as hosts: heritage and wilderness tourism in Nunavut, in M.F. Price (ed.) *People and Tourism in Fragile Environments*, John Wiley, Chichester, 33–50.

Smith, V. and Eadington, W.R. (eds) (1992) *Tourism Alternatives: Potentials and Problems in the Development of Tourism*, University of Pennsylvania Press, Philadelphia.

Social Trends (1996) HMSO, London.

Sofield, T.H.B. (1993) Indigenous tourism development, *Annals of Tourism Research* 20, 729–750.

Sofield, T.H.B. (1996) Anuha Island Resort, Solomon Islands: a case study of failure, in R. Butler and T. Hinch (eds) *Tourism and Indigenous Peoples*, International Thomson Business Press, London, 176–202.

Sooaemalelagi, L., Hunter, S. and Brown, S. (1996) Alternate tourism development in Western Samoa. Paper presented at 3rd Pacific Indigenous Business Conference, 11–13 December 1996, Oxford Koala Hotel, Sydney.

South West Coast Path Steering Group (1996) *Full Report on the Management and Marketing of the South West Coast Path*, South West Coast Path Steering Group, Exeter.

Stabler, M.J. (ed.) (1997) *Tourism and Sustainability. Principles to Practice*, CAB International, Wallingford.

Stankey, G.H., Cole, D.N., Lucas, R.C., Petersen, M.E. and Frissell, S.S. (1985) *The Limits of Acceptable Change (LAC) System for Wilderness Planning*, USDA Forest Service General Technical Report INT-176, Intermountain Forest and Range Experiment Station, Utah.

Stankey, G.H., McCool, S.F. and Stokes, G.L. (1984) Limits of acceptable change: a new framework for managing the Bob Marshall Wilderness complex, *Western Wildlands* 103(3), 33–37.

Stansfield, C.A. (1964) A note on the urban–nonurban imbalance in American recreational research, *Tourist Review* 19(4)/20(1), 196–200, 221–223.

Stansfield, C.A. and Rickert, J.E. (1970) The recreational business district, *Journal of Leisure Research* 2(3), 213–225.

Statistics Canada (1987) *Population and Dwelling Characteristics – Census Divisions and Subdivisions: Quebec*, Minister of Supply and Services Canada, Ottawa.

Statistics Canada (1991) *Population and Dwelling Characteristics – Census Divisions and Subdivisions: Quebec*, Minister of Supply and Services Canada, Ottawa.

Stewart, S. (1996) Bride & gloom, *The Weekend Australian, The Weekend Review: Travel*, 16–17 November, 15–16.

Stoddard, P.H. (1993) Community theory: new perspectives for the 1990s, *Journal of Applied Social Sciences* 1(1), 13–29.

Stoddart, D.R. (1981) Ideas and interpretation in the history of geography, in D.R. Stoddart (ed.) *Geography, Ideology and Social Concern*, Blackwell, Oxford, 1–7.

Stoffle, R.W., Last, C. and Evans, M. (1979) Reservation-based tourism: implications of tourist attitudes for Native American economic development. *Human Organization* 38(3), 300–306.

Swain, D.C. (1970) *Wilderness Defender: Horace M. Albright and Conservation*, University of Chicago Press, Chicago.

Taylor, G. (1995) The community approach: does it really work? *Tourism Management* 16(7), 487–489.

Taylor, G. and Stanley, D. (1992) Tourism, sustainable development and the environment: an agenda for research, *Journal of Travel Research* 31(1), 66–67.

Taylor, M. (1995) *Environmental Change: Industry, Power and Policy*, Avebury, Aldershot.

Teague, P. (1990) The political economy of the regulation school and the flexible specialisation scenario, *Journal of Economic Studies*, 17(5), 32–54.

The Economist (1993) Trekking and wrecking in Nepal, *The Economist*, 35.

Thompson, D. and Wilson, M.J. (1994) Environmental auditing: theory and applications, *Environmental Management* 18(4), 605–615.

Thoreau, H.D. (1968) *Walden*, Everyman Library, Dent, London.

Tourism Concern/World Wildlife Fund (1992) *Beyond the Green Horizon: Principles for Sustainable Tourism*, Tourism Concern, Godalming.

Tourism Policy Group (1995) *Tourism Research Bibliography 1985–Mid 1994*, Ministry of Commerce, Wellington.

Tourism Policy Group (1996) *Tourism Research Bibliography Mid 1994–1996*, Ministry of Commerce, Wellington.

Transport Canada (1988) Air and Maritime Transportation Statistics, *Transport Canada*, Ottawa.

Tuan, Y.-F. (1979) *Landscapes of Fear*, Pantheon, New York.

Turner, F.J. (1893) *The Significance of the Frontier in American History*, Academic Reprints, El Paso, Texas.

Turner, F.J. (1920) *The Frontier in American History*, Henry Holt, New York.

Turner, R.K. (1993a) *Sustainable Environmental Economics and Management: Principles and Practice*, John Wiley, Chichester.

Turner, R.K. (1993b) Sustainability: Principles and Practice, in R.K. Turner (ed.) *Sustainable Environmental Economics and Management: Principles and Practice*, John Wiley, Chichester, 3–36.

US Department of Agriculture (USDA) Forest Service, US Department of the Interior, Bureau of Land Management (1994) *Record of Decision for Amendments to Forest Service and Bureau of Land Management Planning Documents Within the Range of the Northern Spotted Owl*, USDA Forest Service, US Department of the Interior, Bureau of Land Management, Washington.

Urry, J. (1990) *The Tourist Gaze: Leisure and Travel in Contemporary Societies*, Sage, London.

Urry, J. (1996) *Consuming Places*, Routledge, London.

USDA Forest Service (1986) *The 1986 ROS (Recreation Opportunity Spectrum) Book*, Government Printing Office, Washington.

Vaillancourt, J. (1995) Sustainable development: a sociologist's view of the definition, origins and implications of the concept, in M.D. Mekta and E. Ouellet (eds) *Environmental Sociology: Theory and Practice*, Captus Press, Toronto, 219–34.

Vaske, J.J., Donnelly, M.P., Doctor, R.M. and Petruzzi, J.P. (1994) *Social Carrying Capacity at the Columbia Icefield: Applying the Visitor Impact Management Framework*, Canadian Heritage Parks Canada, Calgary.

Vincent, G. (1995) Tourism and sustainable development in Grenada (W.I.): Towards a mode of analysis, unpublished PhD thesis, Department of Geography, McGill University, Montreal.

Volkel-Hutchison, C. (1996) Creating awareness: responsible tourism – an issue for anthropologists? Paper presented at Australian Anthropological Society Conference, 2–4 October 1996, Charles Sturt University, Albury.

Wackernagel, M., McIntosh, J., Rees, W.E. and Wollard, R. (1993) *How Big is Our Ecological Footprint? A Handbook for Estimating a Community's Appropriate Carrying Capacity*, a discussion draft prepared for the Task Force on Planning Healthy and Sustainable Communities, University of British Columbia, Vancouver.

Wackernagel, M. and Rees, W.E. (1995) *Our Ecological Footprint*, New Society Publishers, Gabriola Island.

Walker, C. (1995) Word of mouth, *American Demographics* 17(7), 38–41.

Wall, G. (1993) Ecological reserves and protected areas: the challenge of ecotourism. In seminar on the environment of the academic and scientific community of Mexico, Toluca, Mexico: National Association of Mexican Universities and Inter-American Organization for Higher Education.

Wallace, D.R. (1992) *The Quetzal and the Macaw: The Story of Costa Rica's National Parks*, Sierra Club Books, San Francisco.

Walsh, B. (1996) Authenticity and cultural representation: a case study of Maori tourism operators, in C.M. Hall and S. McArthur (eds) *Heritage Management in Australia and New Zealand: The Human Dimension*, Oxford University Press, Melbourne, 202–207.

Walter, J.A. (1982) Social limits to tourism, *Leisure Studies* 1(2), 295–304.

Wells, K. (1996) The cosmic irony of intellectual property and indigenous authenticity, *Culture and Policy* 7(3), 45–68.

Wells, M.P. (1994) Parks tourism in Nepal: reconciling the social and economic opportunities with the ecological and cultural threats, in M. Munasinghe and J. McNeely (eds) *Protected Area Economics and Policy: Linking Conservation and Sustainable Development*, World Bank, Washington, 319–331.

Western, D. (1993) Defining ecotourism, in K. Lindberg and D. Hawkins (eds) *Ecotourism: A Guide for Planners and Managers*, The Ecotourism Society, North Bennington, 7–11.

Wheeller, B. (1993) Sustaining the ego, *Journal of Sustainable Tourism* 1(2), 121–129.

Whelan, T. (1991) Ecotourism and its role in sustainable development, in T. Whelan (ed.) *Nature Tourism*, Island Press, Washington, 3–22.

Wiesmann, U. (1994) *A Concept of Sustainable Resource Use and its Implications for Research in a Dynamic Regional Context*, University of Bern, Institute of Geography, Bern.

Wight, P.A. (1993a) Sustainable ecotourism: balancing economic, environmental and social goals within an ethical framework, *Journal of Tourism Studies* 4(2), 54–66.

Wight, P.A. (1993b) Ecotourism: ethics or eco-sell? *Journal of Travel Research* 29, 40–45.

Wight, P.A. (1994) Limits of acceptable change: a recreational tourism tool for cumulative effects assessment, in A.J. Kennedy (ed.) *Cumulative Effects in Canada: From Concept to Practice.* Papers from the Fifteenth Symposium held by the Alberta Society of Professional Biologists, 13–14 April, Calgary, 159–178.

Wight, P.A. (1995a) Limits of acceptable change, *Environment Network News* 42, 12–14.

Wight, P.A. (1995b) Sustainable ecotourism: balancing economic, environmental and social goals within an ethical framework, *Tourism Recreation Research* 20(1), 5–13.

Wight, P.A. (1996a) North American ecotourists: market profile and trip characteristics, *Journal of Travel Research* 36(4), 2–10.

Wight, P.A. (1996b) Planning for success in sustainable tourism, presentation to 'Plan for Success' National Conference of the Canadian Institute of Planners, Saskatoon, 2–5 June.

Wilkinson, P.F. (1989) Strategies for tourism in island microstates, *Annals of Tourism Research* 16, 153–177.

Williams, A.M. (1995) Capital and the transnationalisation of tourism, in A. Montanari and A.M. Williams (eds) *European Tourism: Regions, Spaces and Restructuring*, John Wiley, Chichester, 163–176.

Williams, A.M. and Montanari, A. (1995) Introduction: tourism and economic restructuring in Europe, in A. Montanari and A.M. Williams (eds) *European Tourism: Regions, Spaces and Restructuring*, John Wiley, Chichester.

Williams, A.M. and Shaw, G. (eds) (1988) *Tourism and Economic Development: Western European Experiences*, Belhaven Press, London.

Williams, A.M. and Shaw, G. (1995) Tourism and regional development: polarization and new forms of production in the United Kingdom, *Tijdschrift voor Economische en Sociale Geografie* 86(1), 50–63.

Williams, A.M. and Shaw, G. (1996) *Tourism, Leisure, Nature Protection and Agri-Tourism: Principles, Partnerships and Practice*, European Partners for the Environment, Brussels, 1–15.

Williams, P.W. and Gill, A. (1991) *Carrying Capacity Management in Tourism Settings: A Tourism Growth Management Process*, prepared for Alberta Tourism, Edmonton.

Wilson, M. (1997) Community–rail partnerships: benefits to local authorities. Paper to conference on 'Community–Rail Partnerships', Huddersfield, 8 April.

Withiam, G. (1996) Word of mouth: still number one, *Cornell Hotel and Restaurant Administration Quarterly* 37(3), 10.

World Commission on Environment and Development (WCED) (the Brundtland Report) (1987) *Our Common Future*, Oxford University Press, London.

World Conservation Union (IUCN), United Nations Environment Programme (UNEP), and the World Wide Fund for Nature (WWF) (1990) *Caring for the World: A Strategy for Sustainability*, second draft. World Conservation Union (IUCN), United Nations Environment Programme (UNEP), and the World Wide Fund for Nature (WWF).

World Tourism and Travel Council (WTTC) homepage, http://www.wttc.org/

World Tourism and Travel Environmental Research Council (WTTERC) *1993 World Travel and Tourism Environment Review*, WTTERC, Oxford.

World Tourism Organisation (WTO) (1997a) Tourism grows faster in 1996, *WTO News* 1(March), 1–2.

World Tourism Organisation (1997b) 1.6 Billion Tourists by 2020, *WTO News* 2(May), 1–2.

World Tourism Organisation (1997c) *Agenda 21 for the Travel and Tourism Industry*, World Tourism Organisation, Madrid.

Worster, D. (1977) *Nature's Economy: A History of Ecological Ideas*, Cambridge University Press, Cambridge.

Wright, D. (trans.) (1957) *Beowulf*, Penguin, Harmondsworth.

Wyss, F. and Keller, L. (1992), Environmentally acceptable air transport: possibilities and parameters of Swissair's 'Oekkobilanz' environmental audit, in W. Pillman and S. Predi (eds) *Strategies for Reducing the Environmental Impact of Tourism*, International Society for Environmental Protection, Vienna.

Yee, J.G. (1992) *Ecotourism Market Survey: A Survey of North American Tour Operators*. The Intelligence Center, Pacific Asia Travel Association, San Francisco.

Young, B. (1989) New Zealand – Maori tourism on the launch pad, *Tourism Management* 10(2), 153–156.

Young, D. (1994) Traditions Nurtured, *Aorangi* 7, 7–9.

Zazueta, F. (1995) *Policy Hits the Ground Running: Participation and Equity in Environmental Decision Making*, World Resources Institute, Washington.

Zeldin, T. (1994) *An Intimate History of Humanity*, Minerva, London.

Zeppel, H. (1995) Authenticity and Iban longhouse tourism in Sarawak, *Borneo Review* 6(2), 109–125.

Zeppel, H. (1996) 'Come share our culture': Aboriginal tourism in the Northern Territory, in *Interpretation in Action*, Interpretation Australia Association, Collingwood, 121–126.

Zeppel, H. (1997) *Ecotourism and Indigenous Peoples*, issues paper for Ecotourism Information Centre, http://lorenz.mur.csu.edu.au/ecotour/

Ziffer, K. (1989) *Ecotourism: The Uneasy Alliance*, Ernst and Young, Washington.

Zukowski, H. (1994) Masters of ceremony, *Westworld*, Summer, 44–50.

Zurick, D.N. (1992) Adventure travel and sustainable tourism in the peripheral economy of Nepal, *Annals of the Association of American Geographers* 82, 608–628.

Zurick, D.N. (1995) *Errant Journeys: Adventure Travel in a Modern Age*, University of Texas Press, Austin.

Indexes

Place Index

Subject Index